AF614746

Methods in Molecular Biology™

For further volumes:
http://www.springer.com/series/7651

Virus-Induced Gene Silencing

Methods and Protocols

Edited by

Annette Becker

Plant Evodevo Group, Justus-Liebig-Universität Gießen, Gießen, Germany

Editor
Annette Becker
Plant Evodevo Group
Justus-Liebig-Universität Gießen
Gießen, Germany

ISSN 1064-3745 ISSN 1940-6029 (electronic)
ISBN 978-1-62703-277-3 ISBN 978-1-62703-278-0 (eBook)
DOI 10.1007/978-1-62703-278-0
Springer New York Heidelberg Dordrecht London

Library of Congress Control Number: 2012956546

Printed on acid-free paper

Humana Press is a brand of Springer
Springer is part of Springer Science+Business Media (www.springer.com)

Preface

Plants are amazing organisms to study; many are stunningly beautiful, others are critical staple foods, some are important sources for pharmaceuticals, and others can help to elucidate molecular mechanisms required for a plant's development and its interactions with the biotic or abiotic environment. Plant research in the post-genomic era can strongly build on high-quality genome sequences of a large number of plant species, many of which being non-model plants including valuable crop species, plants important for their phylogenetic position, secondary metabolites, or other well-studied research questions. However, analysis of gene functions or even complex molecular mechanisms are still characterized in only a handful of plant model organisms such as *Arabidopsis thaliana* (thale cress), *Oryza sativa* (rice), or several tobacco species.

Meanwhile, functional genomics is vastly lagging behind the speed of genome sequencing as high-throughput gene function assays are difficult to design, specifically for non-model plants. Bioinformatics tools are useful for gene identification and annotation but are of limited value for predictions concerning gene functions as gene functions are uncovered best by experimental approaches. However, mutant collections and protocols for stable genetic transformation are costly and extremely time consuming. Virus-induced-gene-silencing (VIGS) is an easy-to-use, fast, and reliable method to achieve down regulation of target gene expression. Based on modified viral genomes, the method has been applied to diverse species across the angiosperm phylogeny and has yielded already important experiment-derived information on gene functions.

This book provides detailed protocols for VIGS experiments in several plant species including model and non-model plants. A series of methods is described to tackle the sometimes transient and variable phenotypes for researchers working in diverse fields interested in developmental biology, plant–pathogen or herbivore interactions, and plant secondary metabolites. Recently developed protocols for VIGS-derived microRNA production in the plant or protein overexpression are also included in this book, and some chapters are devoted to shortly summarize the molecular mechanisms of VIGS action and the vector systems developed so far and on how to change a plant virus into a VIGS vector.

I hope that this book will provide a valuable resource for researchers from diverse fields of plant biology interested in experimental approaches to analyze gene functions. I am grateful to the authors who have contributed their practical expertise, to John Walker, the series editor for his advice and encouragement, and to Tina Stickan for bringing in her organizational talent. All helped to make this book possible.

Gießen, Germany *Annette Becker*

Contents

Contributors

IAN T. BALDWIN • *Department of Molecular Ecology, Max Planck Institute for Chemical Ecology, Jena, Germany*

ANNETTE BECKER • *Plant Evodevo Group, Justus-Liebig-Universität Gießen, Gießen, Germany*

GUSTAVO BONAVENTURE • *Department of Molecular Ecology, Max Planck Institute for Chemical Ecology, Jena, Germany*

ALON CNA'ANI • *The Robert H. Smith Institute of Plant Sciences and Genetics in Agriculture, The Hebrew University of Jerusalem, Rehovot, Israel*

INDRANIL DASGUPTA • *Department of Plant Molecular Biology, University of Delhi South Campus, New Delhi, India*

XIN SHUN DING • *Plant Biology Division, The Samuel Roberts Noble Foundation Inc., Ardmore, OK, USA*

SON T. DINH • *Department of Molecular Ecology, Max Planck Institute for Chemical Ecology, Jena, Germany*

GUNTHER DOEHLEMANN • *Department of Organismic Interaction, Max Planck Institute for terrestrial Microbiology, Marburg, Germany*

SINEAD DREA • *Department of Biology, University of Leuven, Leuven (Heverlee); Laboratory of Molecular Plant Biology, Institute of Botany and Microbiology, Leuven, Belgium*

JOSEFINA-PATRICIA FERNANDEZ-MORENO • *Fruit Genomics and Biotechnology Laboratory, Instituto de Biología Molecular y Celular de Plantas, (CSIC-UPV), Valencia, Spain*

IVAN GALIS • *Department of Molecular Ecology, Max Planck Institute for Chemical Ecology, Jena, Germany*

XIQUAN GAO • *Department of Plant Pathology and Microbiology, Texas A&M University, College Station, USA; Institute for Plant Genomics and Biotechnology, Texas A&M University, College Station, USA*

KLAUS GASE • *Department of Molecular Ecology, Max Planck Institute for Chemical Ecology, Jena, Germany*

KOEN GEUTEN • *Laboratory of Plant Systematics, Institute of Botany and Microbiology, Leuven, Belgium*

ANTONIO GRANELL • *Fruit Genomics and Biotechnology Laboratory, Instituto de Biología Molecular y Celular de Plantas, (CSIC-UPV), Valencia, Spain*

MARKUS HARTL • *Department of Molecular Ecology, Max Planck Institute for Chemical Ecology, Jena, Germany*

CHRISTIAN HETTENHAUSEN • *Department of Molecular Ecology, Max Planck Institute for Chemical Ecology, Jena, Germany*

JOHN H. HILL • *Department of Plant Pathology, Iowa State University, Ames, IA, USA*

HAIFENG JIA • *Beijing Key Laboratory for Agricultural Application and New Technique, College of Plant Science and Technology, Beijing University of Agriculture, Beijing, China*

ISABELLE JUPIN • *Laboratoire de Virologie Moléculaire, Institut Jacques Monod, CNRS, UMR 7592, Univ Paris Diderot, Sorbonne Paris Cité, Paris, France*

SOFIA KOURMPETLI • *Department of Biology, University of Leuven, Leuven (Heverlee); Laboratory of Molecular Plant Biology, Institute of Botany and Microbiology, Leuven, Belgium*
ELENA M. KRAMER • *Department of Organismic and Evolutionary Biology, Harvard University, Cambridge, MA, USA*
YIZHEN LAI • *School of Life Sciences, Tsinghua University, Beijing, China*
MATTHIAS LANGE • *Plant Evodevo Group, Justus-Liebig-Universität Gießen, Gießen, Germany*
SABRINA LANGE • *School of Geography, Environment and Earth Sciences, University of Wellington, Wellington, New Zealand*
YULE LIU • *School of Life Sciences, Tsinghua University, Beijing, China*
RICHARD S. NELSON • *Plant Biology Division, The Samuel Roberts Noble Foundation Inc., Ardmore, OK, USA*
SVETLANA ORASHAKOVA • *Plant Evodevo Group, University of Bremen, Bremen, Germany*
DIEGO ORZAEZ • *Fruit Genomics and Biotechnology Laboratory, Instituto de Biología Molecular y Celular de Plantas, (CSIC-UPV), Valencia, Spain*
ARUNIMA PURKAYASTHA • *Department of Plant Molecular Biology, University of Delhi South Campus, New Delhi, India*
HEMA RAMANNA • *Plant Biology Division, The Samuel Roberts Noble Foundation Inc., Ardmore, OK, USA*
MEREDITH C. SCHUMAN • *Department of Molecular Ecology, Max Planck Institute for Chemical Ecology, Jena, Germany*
LIBO SHAN • *Department of Plant Pathology and Microbiology, Texas A&M University, College Station, USA; Institute for Plant Genomics and Biotechnology, Texas A&M University, College Station, USA*
BHARTI SHARMA • *Department of Organismic and Evolutionary Biology, Harvard University, Cambridge, MA, USA*
SHWETA SHARMA • *Department of Plant Molecular Biology, University of Delhi South Campus, New Delhi, India*
YUANYUE SHEN • *Beijing Key Laboratory for Agricultural Application and New Technique, College of Plant Science and Technology, Beijing University of Agriculture, Beijing, China*
BEN SPITZER-RIMON • *The Robert H. Smith Institute of Plant Sciences and Genetics in Agriculture, The Hebrew University of Jerusalem, Rehovot, Israel*
YANG TANG • *School of Life Sciences, Tsinghua University, Beijing, China*
DAWIT G. TEKLEYOHANS • *Plant Evodevo Group, Justus-Liebig-Universität Gießen, Gießen, Germany*
ALEXANDER VAINSTEIN • *The Robert H. Smith Institute of Plant Sciences and Genetics in Agriculture, The Hebrew University of Jerusalem, Rehovot, Israel*
KARINA VAN DER LINDE • *Department of Organismic Interaction, Max Planck Institute for terrestrial Microbiology, Marburg, Germany*
DRIES VEKEMANS • *Laboratory of Plant Systematics, Institute of Botany and Microbiology, Leuven, Belgium*
TOM VIAENE • *Laboratory of Plant Systematics, Institute of Botany and Microbiology, Leuven, Belgium*
STEVEN A. WHITHAM • *Department of Plant Pathology, Iowa State University, Ames, IA, USA*
JIANQIANG WU • *Department of Molecular Ecology, Max Planck Institute for Chemical Ecology, Jena, Germany*

NORIKO YAMAGISHI • *Plant Pathology Laboratory, Faculty of Agriculture, Iwate University, Morioka, Japan*
ARAVINDA L. YELLINA • *Plant Evodevo Group, University of Bremen, Bremen, Germany*
NOBUYUKI YOSHIKAWA • *Plant Pathology Laboratory, Faculty of Agriculture, Iwate University, Morioka, Japan*
CHUNQUAN ZHANG • *Department of Plant Pathology, Iowa State University, Ames, IA, USA*

Chapter 1

Virus-Induced Gene Silencing (VIGS) in Plants: An Overview of Target Species and the Virus-Derived Vector Systems

Matthias Lange, Aravinda L. Yellina, Svetlana Orashakova, and Annette Becker

Abstract

The analysis of gene functions in non-model plant species is often hampered by the fact that stable genetic transformation to downregulate gene expression is laborious and time-consuming, or, for some species, even not achievable. Virus-induced gene silencing (VIGS) can serve as an alternative to mutant collections or stable transgenic plants to allow the characterization of gene functions in a wide range of angiosperm species, albeit in a transient way. VIGS vector systems have been developed from both RNA and DNA plant viral sources to specifically silence target genes in plants. VIGS is nowadays widely used in plant genetics for gene knockdown due to its ease of use and the short time required to generating phenotypes. Here, we summarize successfully targeted eudicot and monocot plant species along with their specific VIGS vector systems which are already available for researchers.

Key words: Basal eudicots, Monocots, Rosids, Asterids, VIGS, Gene function analysis

1. Introduction

1.1. A Short Version of the Basic Mechanism of Virus-Induced Gene Silencing

Virus-induced gene silencing (VIGS) exploits the innate plant defense system of posttranscriptional gene silencing (PTGS) against intracellular viral proliferation and extracellular viral movement (1). The modified viral genomes combined with part of the plant's target gene are transformed into the plants via *Agrobacterium tumefaciens* Ti-plasmid-based VIGS. Within the plant cell, transgenic RNA is firstly transcribed and secondly replicated by an endogenous RNA-dependent RNA polymerase (RDRP) enzyme which produces double-stranded RNA molecules triggering PTGS. These double-stranded RNA molecules are recognized by DICER-like

Annette Becker (ed.), *Virus-Induced Gene Silencing: Methods and Protocols*, Methods in Molecular Biology, vol. 975, DOI 10.1007/978-1-62703-278-0_1,

enzymes which cleave the double-stranded RNA into short interfering RNA (siRNA). The short RNA molecules are recognized by the RISC complex, melting the double-stranded siRNAs into single-stranded siRNAs. The RISC complex uses those single-stranded siRNAs to screen RNA populations in the cell for complementary sequences which are then degraded (2, 3). Target sequence-specific siRNA will also be amplified and even transported throughout the plant leading to a systematic gene silencing in distant and often all parts of the plants (4).

1.2. RNA or DNA Plant Viruses Can Elicit VIGS

Several plant viruses have been adopted as VIGS vectors to down-regulate an endogenous plant's target gene after inoculation of the plant with the VIGS vector. Many of these viral vectors are derived from positive-strand RNA viruses such as *Potato virus X* (PVX), *Tobacco mosaic virus* (TMV), and *Tobacco rattle virus* (TRV) which contain an either mono-, bi-, or tripartite genome (5). The VIGS vectors are usually engineered in a way that the modified cDNA copy of the viral genome is inserted into a binary vector system, transformed into an *A. tumefaciens* strain, and used for subsequent agroinoculation. However, alternative methods for delivering the viral vectors to the plant have been developed, and these include mechanical or biolistic inoculation using in vitro synthesized modified viral transcripts or DNA vectors which are delivered directly (5).

A different source of VIGS vectors are the bipartite single-strand DNA viruses and the monopartite DNA viruses that may require a helper virus or satellite DNA in some instances for disease symptom induction. An example is the *Tomato yellow leaf curl China virus* (TYLCCV) with its associated DNAβ satellite which in this case is devoid of the undesired effects of virus infection and instead functions to deliver the sequence to be silenced (5, 6). RNA virus-derived VIGS systems with satellite and helper RNAs have also been developed and used in *Nicotiana tabacum* (tobacco) plants uncoupling viral replication and movement provided by the satellite RNA from gene silencing induction by the helper RNA. This may result in stronger silencing phenotypes compared to the satellite RNA virus alone, also known as satellite virus-induced silencing system (SVISS) (7).

1.3. Viral Cycles of RNA and DNA Viruses

The infection of a plant virus commences after invading a host plant cell with the replication of the viral genome. Replication and transcription of the viral genomes of (+) ssDNA viruses (e.g., geminiviruses) occur in the nucleus and subsequent translation in the cytoplasm (8). Replication, transcription, and translation of (+) ssRNA viruses, which represent the majority of plant viruses modified into successful VIGS vectors, occur in the plasmodesmata-associated cytoplasm via the formation of viral replication complexes (VRC) on host-derived membranes (9).

The replication of mono- and bipartite (+) ssDNA viruses follows two steps: (1) synthesis of the complementary (−) strand by the host plant's enzymes, which results in double-stranded DNA (dsDNA) intermediates, and (2) rolling-circle replication (RCR), where the dsDNA intermediates serve as a template for replication producing ssDNA by virally encoded replication initiation (Rep) proteins (10). The largest genus of the *Geminiviridae* family includes *Begomovirus* with either monopartite or bipartite genomes; its viral cycle is described here in more detail as a representative for ssDNA viruses. The bipartite *Begomovirus* members contain two ssDNA molecules (DNA-A and DNA-B), whereas the monopartite viruses possess only one ssDNA molecule resembling the DNA-A (11). The DNA-A molecule encodes concomitantly all proteins required for successful viral replication, like Rep, REn (replication enhancer protein), TrAP (transcription activator protein), and also the capsid/coat protein CP. The DNA-B codes for a nuclear shuttle protein (NSP) and a movement protein (MP) (10, 12, 13). The NSP is important for nuclear import of the viral genome and thus is crucial for viral replication. Subsequently, the NSP is required for nuclear export of the ssDNA transcripts into the cytoplasm (14). The MP is involved in the local and possibly also in systemic spread of the virus infection.

The *Brome mosaic virus* (BMV) is a member of the genus *Bromoviruses* and is predominantly used as a model virus species for studying replication and translation of (+) ssRNA viruses, and the following viral cycle description is based on studies of BMV. Immediately after the virus enters the plant cell, the viral RNA undergoes initial translation followed by replication employing the viral RDRP. The RNA1 of the BMV genome encodes a monocistronic ORF which is translated into a 109 kDa protein with RNA methylating and capping activity at its N-terminal region and a C-terminal RNA helicase domain. The ORF of RNA2 of the BMV genome contains a protein (94 kDa) with homology to RNA-dependent RNA polymerases. The RNA3 is bicistronic and not required for viral replication but encodes a MP as well as a CP that function in the systemic movement of virus infection. Untranslated *cis*-sequences of the viral genome orchestrate viral gene expression and others are selectively recognized by viral and host factors and incorporate viral RNA it into a VRC (15). The first synthesis of (−) ssRNA intermediates from the original viral genomic (+) ssRNA as well as the following replication steps occur in vesicle-like structures, which are perturbations of internal plasma membranes (16). Reticulon homology proteins (RHPs) are important initial host factors that allow the formation of the VRCs on internal membranes of infected plant cells. In uninfected cells, VRCs induce and stabilize ER membrane tubules (17). Highly complex molecular interactions of viral and host factors also underlie the successful replication and translation of (+) ssRNA viruses (18).

A recent screen in yeast cells for essential genes affecting BMV replication revealed 24 genes whose depleted expression reproducibly inhibited or enhanced BMV RNA replication (19). Among the factors indentified are genes involved in ribosome biosynthesis, cell cycle regulation, as well as the heat shock factor1 (HSF1) or the 26S proteosome components PRE1 and RPT6 (19).

2. Virus Systems Used for Silencing Target Genes in Monocotyledonous Plants

In comparison to the large number of VIGS vector systems adapted for eudicotyledonous plants, only few VIGS systems have been established for monocots, which will be described here only briefly (Table 1). For a more detailed view on monocotyledonous VIGS vector systems, see Remanna et al. Chapter 10 in this book. The *Barley stripe mosaic virus* (BSMV), a tripartite RNA virus, can infect many agronomical important crop species like *Hordeum vulgare* (barley), *Triticum aestivum* (wheat), *Oryza sativa* (rice), and *Zea mays* (maize), and it is used most widely among monocots; for detailed protocols on rice and maize VIGS, see Purkayashtha et al. Chapter 3 and Van der Linde et al. Chapter 4 in this book. BSMV can be used to silence endogenous plant genes in aboveground organs and, principally, also in root tissues. This has been shown in barley and should also be possible in other grass host species; however, not all root genes can be successfully silenced probably due to differential stability of the inserted genes in the BSMV-derived vector (20). The host range of BSMV also includes the recently sequenced cereal model species *Brachypodium distachyon*

Table 1
Monocot species successfully targeted by VIGS

	Viral sources of vectors	References
Brachypodium distachyon	*BSMV*	(20)
Costus spicatus	*BSMV*	(45)
Hordeum vulgare	*BSMV, BMV*	(20, 46, 47)
Oryza sativa	*BMV, RTBV*	(47, 48)
Phalaenopsis spec.	*CymMV*	(49)
Triticum aestivum	*BSMV*	(50)
Zea mays	*BMV*	(47, 51)
Zingiber officinale	*BSMV*	(45)

(purple false brome) in which gene silencing of the *PHYTOENE DESATURASE (PDS)* was successful. Similarly, BSMV has also been applied to *Avena* species (oat), but the silencing effect seems too weak to be of practical use. Besides the Poales, also monocots like the tropical crop *Zingiber officinale* (ginger) and the related plant *Costus spicatus* (Indian head ginger) are susceptible to infection by the BSMV-VIGS system. This is particularly promising as other tropical crops like *Musa acuminata* (banana) and *M. x paradisiaca* (plantain) could possibly similarly be targeted by the well-established BSMV-VIGS system.

An additional grass-derived RNA virus, the BMV, has been adopted for VIGS in barley, rice, and the Va35 cultivar of maize. Also, BMV can be used in maize to study biotrophic interactions, for example, with its native fungal pathogen *Ustilago maydis* by downregulating host genes involved in host–pathogen interactions.

Recently, a more convenient and efficient method using agroinoculation has been developed from the rice-infecting DNA virus *Rice tungro bacilliform virus* (RTBV), allowing for up to 90% downregulation of *PDS* in treated rice leaves.

Finally, a successful VIGS vector system was established for orchids using *Phalaenopsis spec.* as a model cultivar and employing the symptom free *Cymbidium mosaic virus* (CymMV) allowing targeted downregulation of genes in these important horticultural species.

3. Virus Systems Used for Basal Eudicot Species

Several species of basal eudicots were targeted by VIGS approaches (Table 2). Initial experiments suggested that TRV-based vectors efficiently silence *PDS* genes in basal eudicot plants (21). Subsequently, other basal eudicot species, like *Eschscholzia californica* (California poppy, Tekleyohans et al. Chapter 7, this book), *Aquilegia vulgaris* (common columbine, Sharma and Kramer, Chapter 6, this book), *Papaver somniferum* (opium poppy,

Table 2
Basal eudicot species successfully targeted by VIGS

	Viral sources of vectors	References
Aquilegia formosa	*TRV*	(52)
Eschscholzia californica	*TRV*	(53)
Papaver somniferum	*TRV*	(21)
Thalictrum spec.	*TRV*	(54)

Geuten et al. Chapter 5, this book), and *Thalictrum* species, have all been amenable to TRV-VIGS in experiments to functionally analyze floral developmental regulators.

The TRV system consists of two vectors comprising the viral genome, and both are required in a plant cell for infection. The pTRV-RNA1 contains the genes required for viral replication and movement, and the pTRV-RNA2 encodes for the proteins necessary for virion formation. The target gene's partial sequence can be cloned into a multiple cloning site of pTRV-RNA2 which replaces viral proteins not required for VIGS. Both vectors are under control of the *Cauliflower mosaic virus* (CaMV) 35S promoter allowing for constitutive expression in plants (22). Besides its ease of use, there are some additional advantages of the TRV system: viral movement is not excluded from the meristem as in several other viruses, and it allows early plant infection directly below the shoot apical meristem; moreover, genes expressed in floral organs and even pollen can be silenced. TRV does also not produce interfering symptoms in basal eudicot target species upon infection which would otherwise complicate the interpretation of gene-silencing phenotypes (22).

4. Virus Systems Used for Rosids

While VIGS has been developed originally in Solanaceae, several protocols for VIGS systems to silence genes in *Arabidopsis thaliana* (thale cress) have been established as well as for other rosid species (Table 3). For a detailed protocol on VIGS in *A. thaliana*, see Jupin, Chapter 15 in this book. The RNA virus-derived systems of TRV and *Turnip yellow mosaic virus* (TYMV) as well as from the DNA virus *Cabbage leaf curl virus* (CbLCV) have all been employed for targeted gene silencing in *A. thaliana*. VIGS in *A. thaliana* may be the method of choice for several reasons: (1) analysis of redundantly acting genes which are highly similar in sequence by integrating a highly conserved stretch of sequence that is common to all genes into the VIGS vector and (2) genes which have not been targeted by mutagenesis yet or which are embryo- or seedling lethal. In the latter case, VIGS can be employed to downregulate gene expression in later stages of plant development.

Another intriguing possibility in using the VIGS technique might be to co-downregulate reporter transgenes, which help to visually trace the silencing signal within organs and to monitor the silencing effects of metabolite-related genes or their products, an approach successfully applied to tomato fruits (23).

VIGS has recently been used in *Brassica nigra* and *A. thaliana* as a method for investigating oviposition behavior of butterflies on the leaves of their host plants. This is the first description of this

Table 3
Eudicots rosid species successfully targeted by VIGS

	Viral sources of vectors	References
Arabidopsis thaliana	*TRV, TYMV, CbLCV*	(22, 55–57)
Brassica nigra	*TRV*	(58)
Cucumis sativus, C. melo, C. pepo, Citrullus lanatus, Luffa cylindrical, Lagenaria siceraria (Cucurbitaceae)	*ALSV*	(59)
Fragaria ananassa	*TRV*	(60)
Glycine max	*ALSV, BPMV, CMV*	(59, 61–63)
Gossypium hirsutum	*CLCrV*	(64)
Lathyrus odorata	*PEBV*	(65)
Malus domestica	*ALSV*	(28)
Manihot esculenta	*ACMV*	(66)
Medicago truncatula	*PEBV, SHMV*	(65, 67)
Phaseolus vulgaris	*BPMV*	(63)
Pisum sativum	*ALSV, PEBV*	(59, 68)
Pyrus ssp.	*ALSV*	(28)
Rosa hybrida	*TRV*	(35)
Vigna unguiculata, V. angularis	*ALSV*	(59)
Vitis vinifera	*GVA*	(34)

technique used to study ecological aspects of insect–plant interactions in Brassicaceae. For readers specifically interested in using VIGS to study plant–arthropod interactions, please refer to the comprehensive review by Stratmann and Hind (24).

Effective gene silencing can also be accomplished in Cucurbitaceae by the use of *Apple latent spherical virus* (ALSV)-derived vector systems. This particular virus is able to infect a wider range of host plant species and has successfully been expanded to silence endogenous genes also in Fabaceae as proven in *Glycine max* (soybean), *Pisum sativum* (pea), and two *Vigna* species (*V. unguiculata* and *V. angularis*).

VIGS experiments in soybean allow a choice of alternative vector systems based on either *Bean pod mottle virus* (BPMV) or *Cucumber mosaic virus* (CMV). In 2009, a DNA virus copy derived from BPMV was made to increase the efficiency of this system for use in soybean (25). This second-generation BPMV-derived vector was again improved to allow for more sophisticated approaches such as simultaneous silencing of several target genes in soybean and also in *Phaseolus vulgaris* (common bean).

For the important crop plant pea which is difficult to transform transgenically, an additional VIGS system is available which was derived from the tobravirus *Pea early browning virus* (PEBV), a close relative to TRV (26). PEBV-based VIGS systems work also well in *Medicago truncatula* (barrel medic) and *Lathyrus odoratus* (sweet pea). Adding another variety of VIGS vectors that might be used in legume species, Várallyay and coworkers (2010) employed the *Sunn-hemp mosaic virus* (SHMV) as an alternative to PEBV for silencing of target genes. In an initial study, M. truncatula was inoculated, but its efficiency was stated as rather low by the authors.

In addition to the legume family, the rosids also bear several agronomically significant species for which tools to analyze gene functions are required. For researchers and breeders working with *Manihot esculenta* (cassava), a specific VIGS system has been developed using the *African cassava mosaic virus* (ACMV) as the viral source. Similarly, VIGS in grapevine (*Vitis vinifera*) is possible by the use of *Grapevine virus A* (GVA)-derived VIGS vectors, a ssRNA virus from the grapevine-specific genus *Vitivirus*. The authors inoculated in vitro micropropagated plantlets with *A. tumefaciens* in a method called agrodrenching in which infection is established via the roots rather than leaves or other aboveground organs usually used for inoculation.

For *Gossypium hirsutum* (cotton), a silencing vector has been developed from the DNA geminivirus *Cotton leaf crumple virus* (CLCrV). Importantly, the CLCrV vector is able to silence genes in the ovule integuments of cotton plants which are the source of the cotton fibers allowing the molecular analysis of this agronomically important trait at the molecular level.

Attempts to silence host genes in the rose cultivar *Rosa hybrida* "Samantha" have been made in which specific genes required for petal growth were downregulated with VIGS using the TRV-based bipartite vector system. However, in this species, TRV experiments seem to suffer from low efficiency of gene silencing. In the reported experiment, only 3% of the analyzed flowers showed signs of an altered phenotype in the petals.

The possibility to silence genes in *Populus* species would be of major scientific interest, since it would be an exceptional opportunity to silence genes in a eudicot tree species whose full genome sequence provides an excellent source to select target genes. In 2005, a VIGS source derived from the *Poplar mosaic virus* (PopMV) has been developed and successfully used to silence transgenes in *Nicotiana* species (27). However, the final proof of whether the system efficiently works in poplar has so far not been produced. On the other hand, crop tree species of the *Rosaceae* family can be targeted by the aforementioned ALSV system. Seedlings of apple, pear, and Japanese pear were inoculated by particle bombardment and showed reproducible knockdown phenotypes of the silenced genes (28).

5. Viral Systems Used for VIGS in Asterids

Several *Nicotiana* and *Solanum* species, as well as *Petunia hybrida*, were targeted by VIGS with a diverse array of RNA- and DNA-derived VIGS systems making the Solanaceae the most widely used plant family for VIGS experiments (Table 4). *Nicotiana benthamiana* was the first plant species for which VIGS was reported. The *PDS* gene of *N. benthamiana* was silenced by using a vector composed of hybrid sequences of both TMV and *Tomato mosaic virus* (ToMV) which resulted in extensive photo bleaching (29). Subsequently, other species from the genus *Nicotiana* were used in experiments to show that VIGS is more widely applicable. Experimental susceptibility to TRV-elicited VIGS was also demonstrated for the important solanaceous crop plant *Capsicum annum* (pepper).

Asterid species outside the Solanaceae are amenable to gene silencing by VIGS. Recently, it was possible to apply the CMV-based system to silence a floral regulatory gene in the model plant *Antirrhinum majus*. This is particularly important because snapdragon is a long-standing model plant to study genes involved in flower and leaf development, and reverse genetic tools for manipulating gene expression were restricted to tissue culture using shoot

Table 4
Eudicot asterid species successfully targeted by VIGS

	Viral sources of vectors	References
Antirrhinum majus	*CMV*	(69)
Capsicum annum	*TRV*	(70)
Catharanthus roseus	*TRV*	(71)
Nicotiana benthamiana, *N. tabacum*, *Nicotiana* ssp.	*TMV, ToMV, TRV, TLCV, PVX, TGMV, TYLCCNV, ALSV*	(6, 22, 29, 59, 72–74)
Solanum lycopersicon, *S. tuberosum*, *S. bulbocastanum*	*TRV, ASLV*	(59, 75, 76)
Petunia hybrida	*TYDV, PVX, TRV*	(73, 77, 78)
Vaccinium myrtillus	*TRV*	(79)

transformation and regeneration (30). Anthocyanin accumulation in fruits has been investigated in *Vaccinium myrtillus* (bilberry) by injecting the *A. tumefaciens* cultures directly into fruits at a medium developmental stage. Another important application of the VIGS technique comes from *Catharanthus roseus*, the Madagascar periwinkle, where the molecular basis of anticancer alkaloid metabolism is intended to be deciphered with TRV-based VIGS.

6. Virus System Used for Other Dicots

The *Beet curly top virus* (BCTV), a member of the *Geminiviridae*, has been used for VIGS in *Spinacia oleracea* (spinach). This is the only genus of the core eudicot lineage of Caryophyllales for which data on VIGS experiments have been published (31).

7. Agronomically and Scientifically Important Plants Without VIGS Systems

VIGS is still awaiting its application in many plant species important for agriculture and forestry, such as *Allium cepa* (onion), *Helianthus annuus* (sunflower), *Ricinus communis* (castor oil plant), *Arachis hypogaea* (peanut), *Eucalyptus grandis* (flooded or rose gum), and several other tree species. All these plant species' genomes have been sequenced, or a large collection of ESTs are available from which appropriate target gene sequences could be chosen (32).

VIGS is a method to fast and transiently downregulate target gene expression, and with only few exceptions, the targeted gene silencing will not be transferred to subsequent generations. These experimental limitations require initial experiments for each new species or specific tissues targeted, and the optimal time point and location of infection have to be determined experimentally. Problems may arise in many slowly growing perennial or tree species with slowly developing tissues or organs. For example, gene silencing in flower tissues might only be possible when the consequences of target gene VIGS can be observed in a time frame that does not exceed the stability of the VIGS vector (33). Such practical obstacles could have resulted in the only very limited number of tree species successfully targeted by VIGS (28, 34). Rosaceae, apple, and pear trees were shown to be susceptible to ALSV VIGS, and also a cultivar of rose (28, 35).

Major groups of angiosperms for which no VIGS system has been developed to date are the basal monocots such as *Asparagus* species as well as the wealth of the magnoliid clade with prominent crops such as *Cinnamomum verum* (cinnamon) and *Persea americana* (avocado) (36).

In order to understand the genetics of agricultural-related traits, a method of gene downregulation like VIGS is of great importance and could influence breeding strategies. In avocado, the Floral Genome Project (FGP) has provided a wealth of EST data upon which VIGS experiments could be based (37, 38). However, with respect to flower and fruit development, VIGS would be especially useful in conjunction with other experimental procedures such as grafting, where VIGS-treated buds or branches are grafted onto nurse branches as avocado is a tree with a long generation time (39).

Also the basal angiosperm lineages of Nymphaeales, Austrobaileyales, and Amborellales lack a system for efficient downregulation of target genes (32). These groups of plants are scientifically very important as they all have diverged early in angiosperm history and the systematic investigation of gene functions would help to answer important questions about the early radiation of flowering plants and the variability of early angiosperm floral architecture. Many developmental genes, for example, show a remarkable degree of functional conservation among the higher angiosperm model plants, but the ancestral functions of these genes are mostly unknown due to a lack of functional data (40). Establishment of VIGS in these phylogenetically important species would provide a great leap forward toward understanding the enigma of flower evolution, particularly because these species are often recalcitrant to stable genetic transformation (41).

Finally, for gymnosperm species, VIGS experiments have not been reported yet. Gymnosperms comprise many important forestry species and are the sister clade of the flowering plants. Thus, experimental approaches to unravel gene functions are highly desired in the field of gymnosperm genetics (42). Unfortunately, in gymnosperms it is even more difficult to establish VIGS mainly due to a lack of well-characterized viral sources on which the development of VIGS vectors could be based (43), and most angiosperm viruses might be unable to infect gymnosperms.

Since the advent of VIGS around 15 years ago using hybrid tobamoviruses to silence *PDS* in *N. benthamiana*, plant geneticists have greatly benefitted from this new technique, particularly those working with plant species for which genetic transformation is difficult to achieve. Both the number of plant species targeted and the viral sources engineered into VIGS vectors have continued to increase since the advent of VIGS. This development will likely proceed in the future, and one challenge ahead is to establish VIGS for agronomic and phylogenetically important species that lack means to otherwise downregulate gene expression specifically. Additional challenges will be the improvement of protocols for the use of VIGS in already targeted plant species to achieve a higher efficiency and possibly a wider use of VIGS-induced non-transgenic heritable gene expression changes in plants (44).

References

1. Baulcombe D (1999) Viruses and gene silencing in plants. Arch Virol Suppl 15:189–201
2. Ding S-W, Voinnet O (2007) Antiviral immunity directed by small RNAs. Cell 130(3):413–426
3. Waterhouse PM, Fusaro AF (2006) Plant science. Viruses face a double defense by plant small RNAs. Science 313(5783):54–55
4. Kalantidis K et al (2008) RNA silencing movement in plants. Biol Cell 100(1):13–26
5. Purkayastha A, Dasgupta I (2009) Virus-induced gene silencing: a versatile tool for discovery of gene functions in plants. Plant Physiol Biochem 47(11–12):967–976
6. Tao X, Zhou X (2004) A modified viral satellite DNA that suppresses gene expression in plants. Plant J 38(5):850–860
7. Gossele V et al (2002) SVISS—a novel transient gene silencing system for gene function discovery and validation in tobacco plants. Plant J 32(5):859–866
8. Jeske H (2007) Replication of geminiviruses and the use of rolling circle amplification for their diagnosis. In: Czosnek H (ed) Tomato yellow leaf curl virus disease. Springer, Netherlands, pp 141–156
9. Asurmendi S et al (2004) Coat protein regulates formation of replication complexes during tobacco mosaic virus infection. Proc Natl Acad Sci USA 101(5):1415–1420
10. Gutierrez C (2002) Strategies for geminivirus DNA replication and cell cycle interference. Physiol Mol Plant Pathol 60(5):219–230
11. Ha C et al (2008) Molecular characterization of begomoviruses and DNA satellites from Vietnam: additional evidence that the New World geminiviruses were present in the Old World prior to continental separation. J Gen Virol 89:312–326
12. Gutierrez C (2000) DNA replication and cell cycle in plants: learning from geminiviruses. EMBO J 19(5):792–799
13. Gutierrez C (1999) Geminivirus DNA replication. Cell Mol Life Sci 56(3–4):313–329
14. Krichevsky A et al (2006) Nuclear import and export of plant virus proteins and genomes. Mol Plant Pathol 7(2):131–146
15. Noueiry AO, Ahlquist P (2003) Brome mosaic virus RNA replication: revealing the role of the host in RNA virus replication. Annu Rev Phytopathol 41:77–98
16. Rao ALN (2006) Genome packaging by spherical plant RNA viruses. Annu Rev Phytopathol 44:61–87
17. Diaz A, Wang X, Ahlquist P (2010) Membrane-shaping host reticulon proteins play crucial roles in viral RNA replication compartment formation and function. Proc Natl Acad Sci USA 107(37):16291–16296
18. Thivierge K et al (2005) Plant virus RNAs. Coordinated recruitment of conserved host functions by (+) ssRNA viruses during early infection events. Plant Physiol 138(4):1822–1827
19. Gancarz BL et al (2011) Systematic identification of novel, essential host genes affecting bromovirus RNA replication. PLoS One 6(8):e23988
20. Pacak A et al (2010) Investigations of barley stripe mosaic virus as a gene silencing vector in barley roots and in Brachypodium distachyon and oat. Plant Methods 6:26
21. Hileman LC et al (2005) Virus-induced gene silencing is an effective tool for assaying gene function in the basal eudicot species Papaver somniferum (opium poppy). Plant J 44(2):334–341
22. Ratcliff F, Martin-Hernandez AM, Baulcombe DC (2001) Tobacco rattle virus as a vector for analysis of gene function by silencing. Plant J 25(2):237–245
23. Orzaez D et al (2009) A visual reporter system for virus-induced gene silencing in tomato fruit based on anthocyanin accumulation. Plant Physiol 150(3):1122–1134
24. Stratmann JW, Hind SR (2011) Gene silencing goes viral and uncovers the private life of plants. Entomol Exp Appl 140(2):91–102
25. Zhang C et al (2009) Development and use of an efficient DNA-based viral gene silencing vector for soybean. Mol Plant Microbe Interact 22(2):123–131
26. Udvardi MK (2001) Legume models strut their stuff. Mol Plant Microbe Interact 14(1):6–9
27. Naylor M et al (2005) Construction and properties of a gene-silencing vector based on Poplar mosaic virus (genus Carlavirus). J Virol Methods 124(1–2):27–36
28. Sasaki S, Yamagishi N, Yoshikawa N (2011) Efficient virus-induced gene silencing in apple, pear and Japanese pear using Apple latent spherical virus vectors. Plant Methods 7:15
29. Kumagai MH et al (1995) Cytoplasmic inhibition of carotenoid biosynthesis with virus-derived RNA. Proc Natl Acad Sci USA 92(5):1679–1683
30. Cui ML et al (2004) A rapid Agrobacterium-mediated transformation of Antirrhinum majus L. by using direct shoot regeneration from hypocotyl explants. Plant Sci 166(4):873–879
31. Sather DN, Jovanovic M, Golenberg EM (2010) Functional analysis of B and C class floral organ genes in spinach demonstrates their role in sexual dimorphism. BMC Plant Biol 10:46

32. Becker A, Lange M (2010) VIGS—genomics goes functional. Trends Plant Sci 15(1):1–4
33. Folimonov AS et al (2007) A stable RNA virus-based vector for citrus trees. Virology 368(1):205–216
34. Muruganantham M et al (2009) Grapevine virus A-mediated gene silencing in Nicotiana benthamiana and Vitis vinifera. J Virol Methods 155(2):167–174
35. Ma N et al (2008) Rh-PIP2;1, a rose aquaporin gene, is involved in ethylene-regulated petal expansion. Plant Physiol 148(2):894–907
36. Soltis DE et al (2011) Angiosperm phylogeny: 17 genes, 640 taxa. Am J Bot 98(4):704–730
37. Chanderbali AS et al (2009) Transcriptional signatures of ancient floral developmental genetics in avocado (Persea americana; Lauraceae). Proc Natl Acad Sci USA 106(22):8929–8934
38. Wall PK et al (2009) Comparison of next generation sequencing technologies for transcriptome characterization. BMC Genomics 10:347
39. Kasai A et al (2011) Graft-transmitted siRNA signal from the root induces visual manifestation of endogenous post-transcriptional gene silencing in the scion. PLoS One 6:2, e16895
40. Soltis PS et al (2009) Floral variation and floral genetics in basal angiosperms. Am J Bot 96(1):110–128
41. Frohlich MW, Chase MW (2007) After a dozen years of progress the origin of angiosperms is still a great mystery. Nature 450(7173):1184–1189
42. Floyd S, Bowman J (2007) The ancestral developmental tool Kit of land plants. Int J Plant Sci 168(1):1–35
43. Veliceasa D et al (2006) Searching for a new putative cryptic virus in pinus sylvestris L. Virus Genes 32(2):177–186
44. Senthil-Kumar M, Mysore KS (2011) New dimensions for VIGS in plant functional genomics. Trends Plant Sci 16(12):656–665
45. Renner T et al (2009) Virus-induced gene silencing in the culinary ginger (Zingiber officinale): an effective mechanism for down-regulating gene expression in tropical monocots. Mol Plant 2(5):1084–1094
46. Holzberg S et al (2002) Barley stripe mosaic virus-induced gene silencing in a monocot plant. Plant J 30(3):315–327
47. Ding XS et al (2006) Characterization of a brome mosaic virus strain and its use as a vector for gene silencing in monocotyledonous hosts. Mol Plant Microbe Interact 19(11):1229–1239
48. Purkayastha A et al (2010) Virus-induced gene silencing in rice using a vector derived from a DNA virus. Planta 232(6):1531–1540
49. Lu H-C et al (2007) Strategies for functional validation of genes involved in reproductive stages of orchids. Plant Physiol 143(2):558–569
50. Scofield SR et al (2005) Development of a virus-induced gene-silencing system for hexaploid wheat and its use in functional analysis of the Lr21-mediated leaf rust resistance pathway. Plant Physiol 138(4):2165–2173
51. van der Linde K et al (2010) Systemic virus-induced gene silencing allows functional characterization of maize genes during biotrophic interaction with Ustilago maydis. New Phytol 189(2):471–483
52. Kramer EM et al (2007) Elaboration of B gene function to include the identity of novel floral organs in the lower eudicot Aquilegia (Ranunculaceae). Plant Cell 19:756–766
53. Wege S et al (2007) Highly efficient virus-induced gene silencing (VIGS) in California poppy (Eschscholzia californica): an evaluation of VIGS as a strategy to obtain functional data from non-model plants. Ann Bot 100(3):641–649
54. Di Stilio VS et al (2010) Virus-induced gene silencing as a tool for comparative functional studies in Thalictrum. PLoS One 5(8):e12064
55. Burch-Smith TM et al (2006) Efficient virus-induced gene silencing in Arabidopsis. Plant Physiol 142(1):21–27
56. Pflieger S et al (2008) Efficient virus-induced gene silencing in Arabidopsis using a 'one-step' TYMV-derived vector. Plant J 56(4):678–690
57. Turnage MA et al (2002) Geminivirus-based vectors for gene silencing in Arabidopsis. Plant J 30(1):107–114
58. Zheng SJ et al (2010) Disruption of plant carotenoid biosynthesis through virus-induced gene silencing affects oviposition behaviour of the butterfly Pieris rapae. New Phytol 186(3):733–745
59. Igarashi A et al (2009) Apple latent spherical virus vectors for reliable and effective virus-induced gene silencing among a broad range of plants including tobacco, tomato, Arabidopsis thaliana, cucurbits, and legumes. Virology 386(2):407–416
60. Chai Y-M et al (2011) FaPYR1 is involved in strawberry fruit ripening. J Exp Bot 62(14):5079–5089
61. Zhang C, Ghabrial SA (2006) Development of bean pod mottle virus-based vectors for stable protein expression and sequence-specific virus-induced gene silencing in soybean. Virology 344(2):401–411
62. Nagamatsu A et al (2007) Functional analysis of soybean genes involved in flavonoid

biosynthesis by virus-induced gene silencing. Plant Biotechnol J 9(3):348–358
63. Zhang C et al (2010) The development of an efficient multipurpose bean pod mottle virus viral vector set for foreign gene expression and RNA silencing. Plant Physiol 153(1):52–65
64. Tuttle JR et al (2008) Geminivirus-mediated gene silencing from cotton leaf crumple virus is enhanced by low temperature in cotton. Plant Physiol 148(1):41–50
65. Grønlund M et al (2008) Virus-induced gene silencing in Medicago truncatula and Lathyrus odorata. Virus Res 135(2):345–349
66. Fofana IBF et al (2004) A geminivirus-induced gene silencing system for gene function validation in cassava. Plant Mol Biol 56(4):613–624
67. Varallyay E et al (2010) Development of a virus induced gene silencing vector from a legume infecting tobamovirus. Acta Biol Hung 61(4):457–469
68. Constantin GD et al (2004) Virus-induced gene silencing as a tool for functional genomics in a legume species. Plant J 40(4):622–631
69. Kim BM, Inaba J, Masuta C (2011) Virus induced gene silencing in Antirrhinum majus using the Cucumber mosaic virus vector: functional analysis of the AINTEGUMENTA (Am-ANT) gene of A. majus. Hort Environ Biotechnol 52(2):176–182
70. Chung E et al (2004) A method of high frequency virus-induced gene silencing in chili pepper (Capsicum annuum L. cv. Bukang). Mol Cells 17(2):377–380
71. Liscombe DK, O'Connor SE (2011) A virus-induced gene silencing approach to understanding alkaloid metabolism in Catharanthus roseus. Phytochemistry 72(16):1969–1977
72. Li D et al (2008) Tomato leaf curl virus satellite DNA as a gene silencing vector activated by helper virus infection. Virus Res 136(1–2):30–34
73. Ruiz MT, Voinnet O, Baulcombe DC (1998) Initiation and maintenance of virus-induced gene silencing. Plant Cell 10(6):937–946
74. Kjemtrup S et al (1998) Gene silencing from plant DNA carried by a Geminivirus. Plant J 14:91–100
75. Brigneti G et al (2004) Virus-induced gene silencing in Solanum species. Plant J 39(2):264–272
76. Liu Y, Schiff M, Dinesh-Kumar S (2002) Virus-induced gene silencing in tomato. Plant J 31:777–786
77. Atkinson RG et al (1998) Post-transcriptional silencing of chalcone synthase in petunia using a geminivirus-based episomal vector. Plant J 15(5):593–604
78. Chen J-C et al (2004) Chalcone synthase as a reporter in virus-induced gene silencing studies of flower senescence. Plant Mol Biol 55(4):521–530
79. Jaakola L et al (2010) A SQUAMOSA MADS box gene involved in the regulation of anthocyanin accumulation in bilberry fruits. Plant Physiol 153(4):1619–1629

Chapter 2

Rationale for Developing New Virus Vectors to Analyze Gene Function in Grasses Through Virus-Induced Gene Silencing

Hema Ramanna, Xin Shun Ding, and Richard S. Nelson

Abstract

The exploding availability of genome and EST-based sequences from grasses requires a technology that allows rapid functional analysis of the multitude of genes that these resources provide. There are several techniques available to determine a gene's function. For gene knockdown studies, silencing through RNAi is a powerful tool. Gene silencing can be accomplished through stable transformation or transient expression of a fragment of a target gene sequence. Stable transformation in rice, maize, and a few other species, although routine, remains a relatively low-throughput process. Transformation in other grass species is difficult and labor-intensive. Therefore, transient gene silencing methods including *Agrobacterium*-mediated and virus-induced gene silencing (VIGS) have great potential for researchers studying gene function in grasses. VIGS in grasses already has been used to determine the function of genes during pathogen challenge and plant development. It also can be used in moderate-throughput reverse genetics screens to determine gene function. However, the number of viruses modified to serve as silencing vectors in grasses is limited, and the silencing phenotype induced by these vectors is not optimal: the phenotype being transient and with moderate penetration throughout the tissue. Here, we review the most recent information available for VIGS in grasses and summarize the strengths and weaknesses in current virus–grass host systems. We describe ways to improve current virus vectors and the potential of other grass-infecting viruses for VIGS studies. This work is necessary because VIGS for the foreseeable future remains a higher throughput and more rapid system to evaluate gene function than stable transformation.

Key words: VIGS, Monocotyledons, RNAi, Plant viruses, Barley, Wheat, Maize, *Brachypodium*, *Setaria*

1. Introduction

Genome and EST sequencing have produced massive amounts of sequence information in the plant genomic era. Genome sequences of many grasses are available (1–4). However, the functions of many of the genes in these sequences are unknown, and determining

Annette Becker (ed.), *Virus-Induced Gene Silencing: Methods and Protocols*, Methods in Molecular Biology, vol. 975,
DOI 10.1007/978-1-62703-278-0_2, © Springer Science+Business Media New York 2013

their functions is a major challenge for plant biologists studying these species. Gene function can be studied either by overexpression, stable gene disruption (knockout), or silencing (knockdown) procedures (5, 6). Chemical or irradiation mutagenesis, insertional mutagenesis, RNAi, and virus-induced gene silencing (VIGS) are some of techniques used to achieve gene knockout or knockdown in both monocotyledons and dicotyledons. Chemical, irradiation, and insertion mutagenesis methods are extensively used to study the function of single genes; however, they generally do not allow the analysis of gene families (7–10). Methods exist to produce stably modified plants where gene families are silenced by targeting conserved sequence domains. However, these methods involve stable transformation which for grass species is a long, if at all achievable, process. VIGS is a rapid alternative knockdown method that allows silencing of individual genes or gene families to study their function in plant development, disease resistance, and during abiotic stress (5, 11–14). As generally occurs with stable RNAi, VIGS involves sequence-specific RNA degradation at the posttranscriptional level in plants (15), but unlike stable RNAi, results can be obtained within 2 months from target identification. Approximately 37 VIGS vectors are available for studies with dicotyledonous plants, and over 37 dicotyledonous species have been studied with these vectors. Findings from this research have resulted in a greater understanding of vegetative and reproductive plant development, biotic and abiotic stress tolerance, and nodule development (11).

For grass species, fewer VIGS vectors are available and thus fewer species (11 crop plants) have been studied (16). Although VIGS has been applied in grass species to study genes involved in biotic stress tolerance and cell wall biosynthesis (16, 17), there are some significant limitations apparent in these studies compared with those involving dicotyledonous plants (particularly those involving *Nicotiana benthamiana*). Highest among these is the loss of the silencing phenotype with time as the plant develops. In this review, we will provide information on the VIGS vectors available for functional genomic studies in grasses, some breakthrough findings using these vectors, the current limitations in VIGS studies in grasses, and ways to improve this situation both by modifying current vectors and identifying new vectors.

2. VIGS Vectors Available for Grasses

In the past 9 years, three VIGS vectors have been developed for grass species: *Barley stripe mosaic virus* (BSMV; (18–20)), *Brome mosaic virus* (BMV; (21, 22)), and *Rice tungro bacilliform virus*

(RTBV; (23)). A brief description of each vector and their inoculation protocols follows:

2.1. BSMV-VIGS Vector

BSMV is a positive-strand RNA virus of the genus *Hordeivirus*. It has a tripartite genome composed of α, β, and γ segments. Holzberg et al. (18) modified a clone of the ND18 strain of BSMV to function as a VIGS vector by deleting the coat protein gene within the plasmid representing the β genomic RNA and adding a multiple-cloning site (*Pac*I and *Not*I) downstream of the γb gene for insertion of foreign gene fragments (120–500 bp). Bruun-Rasmussen et al. (24) created a similar BSMV vector by inserting sequence containing slightly different restriction sites (*Sma*I, *Pac*I, and *Bam*HI) downstream of the γb gene and maintaining the coat protein gene in the β genomic RNA. The inclusion of the coat protein gene decreased undesirable necrosis on infected barley leaves (24, 25). Later, Meng et al. (20) inserted the three BSMV cDNAs from the Holzberg et al. (18) constructs, the β segment containing an active coat protein gene, each into a separate binary vector between a 35S promoter and a ribozyme/nopaline synthase 3′ terminator, the ribozyme added to create a more functional 3′ terminal sequence after transcription of the plasmid *in planta*. A different version of the BSMV vector was developed by modifying the β and γ RNAs of the ND18 strain (19). The start codon for the coat protein in the sequence representing β RNA was mutated as was the start codon of the γb gene in the γ RNA sequence to create a *Bam*HI site for insertion of the foreign gene fragment. VIGS using the BSMV vectors was demonstrated in barley (18), wheat (25), ginger (26), *Haynaldia villosa* (27), *Brachypodium distachyon* (28, 29), and oat (29).

2.1.1. Inoculation Methods

Inoculum for the BSMV vector developed by Holzberg et al. (18) and Tai et al. (19) is produced by in vitro transcription of linearized plasmids containing the BSMV genome sequences. The transcripts are capped during the reactions and the products mixed with FES buffer prior to rub-inoculation of seedlings (18, 25). Meng et al. (20) developed a modified system using particle bombardment of binary vectors containing the BSMV genomes for introduction into barley. The biolistic-based delivery system does not require expensive in vitro transcription enzymes, but initial costs can be extensive if the purchase of a commercial biolistic gene gun and gold particles is required.

2.2. BMV-VIGS Vector

BMV is a positive-strand RNA virus of the genus *Bromovirus*. It has a tripartite genome composed of RNAs 1, 2, and 3. Ding et al. (21) cloned and modified the fescue strain of BMV (F-BMV, isolated from *Festuca arundinacea*) to function as a VIGS vector. A clone of genomic RNA 3 representing the Russian strain of BMV (R-BMV) was used with genomic clones representing RNAs 1 and

2 of F-BMV to allow easy foreign gene insertion (due to a unique *Hin*dIII restriction site in the R-BMV RNA 3 clone) and infectivity to rice (due to the use of clones representing F-BMV RNAs 1 and 2 (21)). Foreign gene fragments were inserted downstream of the coat protein stop codon within the cDNA representing RNA 3. The RNA 3 clone representing F-BMV later was modified by replacing a portion of the intergenic sequence between movement and coat protein genes with a corresponding fragment from the clone representing the R-BMV RNA 3. This was done to increase the accumulation of RNA 3 and the subgenomic RNA 4 from RNA 3, both containing the target host gene fragment that serves as substrate for RNA silencing, during infection (30). The BMV vector containing this chimeric RNA 3 (C-BMV RNA 3), when analyzed against parental viruses, yielded more progeny virus than F-BMV and induced fewer disease symptoms than the R-BMV in rice (21). More recently, foreign gene fragments were directionally inserted into the modified F-BMV RNA 3 clone between newly added *Nco*I and *Avr*II restriction sites immediately downstream of the coat protein sequence, thus yielding a VIGS vector predominantly composed of F-BMV sequence (31). This BMV vector was further modified by cloning sequences representing RNAs 1, 2, and 3 into a modified pCAMBIA 1300 binary vector between a double 35S promoter and a ribozyme to create a more functional 3′ terminal sequence after transcription *in planta* (31). The BMV vector used by Pacak et al. (22) for VIGS was developed from R-BMV (32) for RNA recombination studies. In this vector, the R-BMV RNA 3 clone was modified by replacing the existing sequence 3′ of the coat protein open reading frame (ORF) with two restriction sites separated by a 337-nt spacer, to allow expression of inverted repeats, followed by 295 nt from the 3′ end of wild-type R-BMV RNA 1. VIGS using the BMV vectors was demonstrated in rice, maize, and barley (21) and also is under study in *Setaria italica* (foxtail millet) and *Sorghum bicolor* (sorghum) (H. Ramanna, X.S. Ding and R.S. Nelson, unpublished data).

2.2.1. Inoculation Methods

Inoculum for the older BMV vectors developed by Ding et al. (21, 33) and Pacak et al. (22) is produced by in vitro transcription of linearized plasmids containing the BMV genome sequences. The transcripts are capped during the reaction, and the product is mixed with FES buffer and then rub-inoculated to seedlings (21). We determined that inoculation of the BMV vector to *N. benthamiana* first, as an intermediate host, provides a high titer of virus in extract from these plants for subsequent inoculation to the grass host (33). For the DNA-based BMV vector, an *Agrobacterium*-mediated vacuum infiltration method was developed to introduce the virus into rice (31). Because the current *Agrobacterium*-mediated vacuum infiltration method is not optimized for all grasses, we infiltrate *N. benthamiana* as an intermediate host, using a needle-less syringe, and the sap from infected leaves of this species is rub-inoculated to the target grass plants. Using this method, VIGS has been demonstrated in several grasses (e.g., foxtail millet and sorghum) (H. Ramanna, X.S. Ding and R.S. Nelson, unpublished data). We propose that this method may be useful for a wide range of grasses. Van der Linde et al. (34) further improved VIGS studies using the BMV vector by normalizing, between treatments, the vector inoculum loads obtained from the *N. benthamiana* intermediate host that were destined for the grass host.

2.3. RTBV-VIGS Vector

RTBV is a double-stranded DNA virus of the genus *Pararetrovirus*. It has a genome of approximately 8 kb encoding four ORFs (I to IV). RTBV was modified to serve as a VIGS vector by assembling the viral DNA molecule as a partial dimer within the T-DNA of a binary plasmid. In addition, the RTBV promoter was replaced with the constitutively expressed maize ubiquitin promoter, and a tRNA-binding site and a multiple-cloning site, the latter for the insertion of foreign gene fragments, were added. VIGS using the RTBV vector was demonstrated in rice (23).

2.3.1. Innoculation Method

The RTBV vector is introduced into the host plant through *Agrobacterium*-mediated injection (23).

3. Analysis of Gene Function with Available VIGS Vectors

Currently, BSMV and BMV are the only vectors widely used to characterize gene function in grasses through VIGS. BSMV vectors have been used to study genes involved in disease resistance, such as leaf rust resistance genes *Lr21* (25) and *Lr10* (35) in wheat; stem rust resistance genes *TaRLK1*, *2*, and *3* in wheat (36) and contig4211 of the *NecS1* gene (*NecS1* encodes a cation/proton-exchanging protein (HvCAX1)) in barley (37); powdery mildew resistance genes *HSP90* (38), *Blufensin1* (*Bln1*) (20), and *WRKYs 1* and *2* in barley (39) and serine and threonine protein kinase gene (*Stpk-V*) in wheat (40); necrotrophic fungal resistance gene *ToxA-binding protein1* (*ToxABP1*) in wheat (41); stripe rust fungal resistance genes *TaHsp90.2* and *TaHsp90.3* in wheat (42); and genes involved in nonhost resistance in barley (43). BSMV vectors also have been used to study genes involved in insect resistance in wheat, such as the aphid resistance genes *WRKY53* and *Pal* (44); genes involved in cell wall metabolism in barley, such as *P23k* (45) and *CesA* (17); root genes involved in phosphate acquisition, such as *IPS1*, *PHR1*, and *PHO2* (29); and a gene involved in seedling growth in wheat, *TaHsp90.1* (42).

BMV vectors have been used to study maize genes involved in interactions with the fungus, *Ustilago maydis*, such as *Tps6/11*, *Ecb*

(CD967190), and *Bti* (BM380261). Silencing *Ecb* and *Bti* did not significantly alter leaf colonization by the virus, whereas silencing the *Tps6/11* gene increased tissue susceptibility to *U. maydis* (34). More recently, the BMV vector was used to silence an m-type thioredoxin in maize (46). Tissue silenced for the expression of this gene was more susceptible to a systemic infection by *Sugarcane mosaic virus*, a potyvirus.

4. Limitations of VIGS in Grasses and Approaches to Overcome These Limitations

Despite the significant advances made using BSMV- and BMV-based silencing vectors in grasses, these vectors are not optimized for maximum silencing efficiency in the currently utilized grass species. Also, the usefulness of VIGS in many other grasses has not been studied (12, 16). Below, we discuss the limitations for silencing evident for existing vectors and the absence of a virus vector for other grass species. We also discuss methods to improve the silencing response and the potential of other viruses as VIGS vectors.

For BSMV- and BMV-based silencing vectors, not all cultivars or accessions within a particular host species show measurable target gene silencing (21, 24, 29). This can be due to poor infectivity, replication, or movement of the silencing vector within that host. For barley, the "Black Hulless" cultivar displayed the best silencing phenotype (18, 24). In a separate study using other cultivars, "Spire" and "Clansman" exhibited the best silencing phenotype and reduction in target transcript levels (38). For wheat, the "Bobwhite" cultivar displayed more photobleaching and reduction in target transcript level than "Clark" at 22°C (47). Others saw similar results with "Bobwhite" and six undisclosed hexaploid wheat cultivars (25). None of the surveys were fully comprehensive, so other varieties may exist that provide better silencing. In a significant later study, it was determined that a BSMV vector expressing a fragment of the GFP gene (BSMV-GFP) induced expression of pathogenesis-related and phenylalanine ammonia-lyase genes, all associated with plant defense, in the wheat cultivar Renan (48). These researchers determined that the induction of these genes was correlated with enhanced resistance to *Magnaporthe oryzae*, the blast pathogen, but had no effect on the development of powdery mildew disease induced by *Blumeria graminis*. These findings make it clear that researchers must run appropriate controls to determine the influence of a virus infection itself on host metabolism irrespective of any targeted effects during VIGS.

In a survey of rice varieties, BMV caused modest mosaic symptoms on cultivars IR64, Drew, and Cypress and thus the greatest potential for minimum confounding effects on silencing studies (21). However, no BMV strain was able to infect Nipponbare, one of the most valuable rice cultivars for plant molecular biologists and geneticists (21). Also, BMV induced severe disease symptoms (local necrosis on leaf lamina) on cultivars Moroberekan and IAC165 which would seriously confound the interpretation of findings from silencing studies. Even more serious systemic necrosis systems are induced by BMV on most cultivars of maize, including widely used B73 and Mo17 ((21), X.S. Ding and R.S. Nelson, unpublished data). These cultivars and all others that display systemic necrosis are not suitable hosts for VIGS studies. We include a table listing varieties within grass species and their responses to viruses and viral vectors used in VIGS studies (Table 1).

Although there has been considerable exploration of silencing responses within varieties for their response to the BSMV and BMV silencing vectors, the search among species for silencing responses is less complete. BMV can infect numerous species of *Poaceae* under the greenhouse conditions (49, 50), and further experiments are needed to determine if VIGS can be applied to them, particularly those recalcitrant to stable genetic transformation. One example is the use of the BMV vector to silence genes in *S. italica* (foxtail millet), a model C4 grass closely related to the prospective biofuel crop grass, *Panicum virgatum* (switchgrass) (H. Ramanna, X.S. Ding and R.S. Nelson, unpublished results). BSMV has recently been shown to be suitable for VIGS studies in *B. distachyon*, a model plant for C3 crops such as barley, wheat, and oats (29). However, it is clear from the current studies that expanding the use of the current vectors into a new grass species always will require a survey of varieties and accessions within the new target species to obtain the best silencing phenotypes.

Temperature also influences VIGS phenotypes in plants. In dicotyledonous plants, it has been shown that low temperature enhances the appearance of visible phenotypes during VIGS and higher temperatures enhance the silencing of the virus genome (5, 51–55). It is possible that the lower temperature allows continuous accumulation of the silencing substrate (i.e., the virus vector) to levels which induce gene silencing while higher temperatures lead to destruction of the substrate and no endogenous gene silencing. A similar finding was made with the BSMV vector in wheat where temperatures of 18–22°C yielded better silencing phenotypes and greater target gene transcript reductions than observed at 26°C (47). Also in barley, temperatures of 20–24°C provided better results than 16 or 28°C (24). However, for BMV, Ding et al. (56) reported that under a low temperature condition (24/20°C, day/night), this virus infected and accumulated predominantly in cells near and within vascular cells of barley. Although this study did not involve VIGS, it is important to note that even under conditions that improve gene silencing, the virus spread in the infected plants may be affected leading to an incomplete penetration of the silencing phenotype. Thus,

Table 1
Comparison of species and varietal responses during virus challenge and VIGS

Virus	Plant species	Cultivar, variety, or PI	Response to virus	Response during VIGS	References
BSMV	*Hordeum vulgare* (barley)	Black Hulless	–[a]	Best	(18, 24)
		Golden Promise, Pallas, Ingrid, Chess, Simba, Relief	–	Weak, delayed silencing phenotype and viral symptom	
		Spire, Clansman, Tyne	–	Best	(38)
		Digger	–	Weak silencing and strong virus symptoms in systemic leaves	
		Clansman (*Mla13*), C.I. 16151 (*Mla6*)	–	Best	(20)
		C.I. 16137 (*Mla1*), C.I.16155 (*Mla13*), Sultan-5 (*Mla12*), Golden promise, C.I. 16147 (*Mla7*), C.I.16149 (*Mla10*), HOR11358 (*Mla9*), C.I. 16143 (*Mlk*), C.I. 15229 (Steptoe), Ingrid (*Mlo*), Harrington, C.I. 16139 (*Mlg*), OWB rec, C.I. 16145 (*Mlp*), C.I. 16141 (*Mlh*), *mlo-5* BC_7 Ingrid, C.I. 15773 (Morex)	–	Weak silencing phenotype	
BSMV	*Triticum aestivum* (wheat)	Bobwhite	–	Best	(47)
		Clark	–	Weak silencing phenotype	
		Bobwhite	–	Best compared with six other undisclosed tested varieties	(25)

BSMV	*Avena sativa* (oat)	Belinda	Weak virus infection (compared to barley)	Weak silencing (compared to barley)	(29)
	Avena strigosa (oat)	S75	Weak virus infection (compared to barley)	Weak silencing (compared to barley)	
BMV	*Oryza sativa* (rice)	Drew, IR8, IR64, Cypress, Pokkali, M-202	Mild virus symptom	Only analyzed for IR8 and IR64: Visible, transient silencing	[(21); Ding et al. unpublished data]
		Moroberekan, IAC65	Systemic necrosis	–	
		Nipponbare	No infection		
BMV	*Zea mays* (maize)	Va35	Mild virus symptom	Visible, transient silencing	[(21); Ding et al. unpublished data]
		B73, Mo17, W22	Severe systemic necrosis	–	
BMV	*Setaria italica* (Foxtail millet)	Yugu, German R, PI 212626	Yugu and German R only, tested in detail: mild virus symptom	Best	Ramanna et al. unpublished data
		PI 315088		Weak silencing phenotype	
		PI 391643		No silencing phenotype	
BMV	*Sorghum bicolor* (sorghum)	BTx623 (PI564163)	Mild virus symptom	Best	Ramanna et al. unpublished data
		RioS, PI 651495, PI 651497	–	Weak silencing phenotype	
RTBV	*Oryza sativa* (rice)	TN-1	No stunting or yellowing	Mild, longer lasting but still transient	(23)

[a]–, not determined

temperature effects should always be analyzed when a new plant–virus–vector combination is chosen for studies.

Another factor that may influence silencing phenotypes during VIGS is the orientation of the foreign gene insert. Host gene fragments inserted in the sense orientation within the virus vector are never superior and in some instances are inferior to those inserted in the antisense orientation ((18, 22, 24, 25, 57) Fig. 1). Early studies indicated that use of an inverted repeat target sequence in the BSMV vector increased the visual silencing phenotype, if not the downregulation of the target sequence, compared with an antisense fragment (57). Recent studies however determined that short inverted repeats can be very unstable and less efficient than sense constructs for VIGS in BSMV and BMV (22, 29). Further work is necessary to determine the general usefulness of inverted repeats during silencing in grasses.

As hinted at in the sentence above, foreign gene fragment insert stability in the silencing vector is an important consideration during VIGS studies. Indeed, it is likely the most important factor that limits the effective use of VIGS in grasses and probably in most plant species. Researchers who utilize *Tobacco rattle virus* (TRV) to silence genes in *N. benthamiana* are among the few who can depend on the silencing phenotype to persist in new systemically infected tissue and penetrate all cell types within the involved tissue (58). This phenotype often is associated with the maintenance of the foreign gene fragment in the vector. Similar results were also obtained in soybean plants silenced for their PDS gene using the *Apple latent spherical virus*-based silencing vector (59). VIGS in grasses, however, is always associated with a transient phenotype in systemically infected tissue, usually involving only two to three leaves and the intervening stem tissues (18, 21, 24, 25). In addition, penetration of the visible silencing phenotype in leaves was often incomplete, appearing only as large or narrow stripes between longitudinal vascular bundles. These transient and incomplete visible silencing phenotypes are generally associated with the loss of the foreign gene fragment from the recombinant virus vectors during infection. Bruun-Rasmussen and colleagues studied this carefully and determined that although a larger insert (584 nt of PDS gene in BSMV) induced clear photobleaching in the second leaf above the inoculated leaf, the areal coverage was less than for those plants inoculated with BSMV containing 400 and 275 nt inserts (24). This loss of areal coverage was closely correlated with the loss of the 584 nt insert from the viral genome.

The mechanism driving the loss of a foreign gene fragment from a virus vector is not fully understood. Possible causes include deletion of the insert due to RNA polymerase hopping or recombination between viral RNAs (60–64). A recombined virus, now without an insert, will likely accumulate to very high levels compared with virus still containing an insert. Thus, loss of insert not

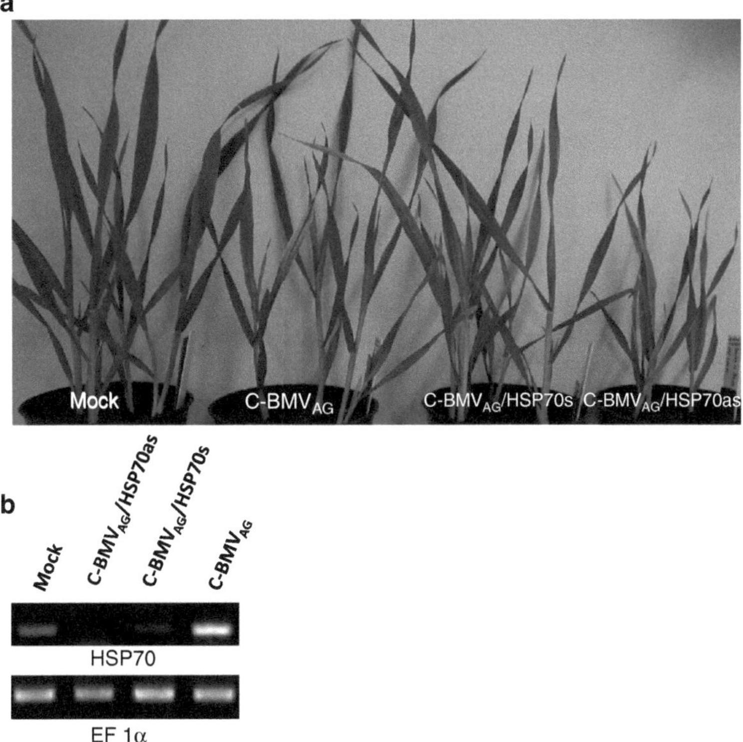

Fig. 1. Virus-induced gene silencing of a heat shock protein 70 (HSP70) was more effective with an antisense than a sense fragment insert in barley using the C-BMV$_{A/G}$ vector. Seedlings of barley cv. Morex were inoculated with C-BMV$_{A/G}$ or C-BMV$_{A/G}$ carrying a sense (s) or antisense (as) HPS70 insert. The inoculated plants were grown inside a growth chamber set at 24/20°C (day/night) and photographed 10 days postinoculation (**a**). The second systemically infected leaves of these plants were harvested and extracted to determine the level of HSP70 transcript through semiquantitative RT-PCR (**b**). The barley elongation factor 1α (EF 1α) gene was used as the internal control. Gel images show PCR products obtained after 30 reaction cycles. (1) Mock, (2) C-BMV$_{AG}$/HSP70as, (3) C-BMV$_{AG}$/HSP70s, and (4) C-BMV$_{AG}$. Note the more severe stunting and greater target transcript knockdown in plants inoculated with C-BMV$_{AG}$/HSP70as.

only removes the fragment responsible for inducing target gene silencing, but yields a virus that will outcompete any remaining virus that retains the insert and possibly lead to confounding virus symptoms. Bruun-Rasmussen et al. (24) observed that the BSMV vector that originally contained a 584 nt insert accumulated to much higher levels than those with 400 and 275 nt inserts, but this higher accumulation was associated with loss of the insert, in agreement with the above assertion. In a later study, this laboratory also determined that a BSMV vector with a 493 nt insert from a different gene was more stable and this was associated with a silencing phenotype (29). Thus, it is not only the length of the insert, but the nucleotide content that influences insert stability.

Virus inoculation procedures often have a significant effect on VIGS studies. In the last few years, an *Agrobacterium*-mediated infiltration procedure to introduce VIGS vectors has emerged as a rapid and reliable tool for gene silencing in *N. benthamiana*

(5, 65–68). Recent improvements of this infiltration procedure allowed researchers to achieve reliable silencing results in more plant species of the *Solanaceae* family, lettuce, *Arabidopsis thaliana* (L.), and also in fruits of tomato (69–72). In grasses, several DNA viruses were previously reported to be successfully introduced into rice, maize, and wheat through an *Agrobacterium*-mediated injection procedure (73–77). In 2009, a first report on an *Agrobacterium*-mediated vacuum infiltration procedure for VIGS in rice appeared (31). These researchers modified the BMV silencing vector by placing the viral sequences behind a doubled 35S promoter within a modified binary vector. Vacuum infiltration of rice cultivar IR64 with a solution of *Agrobacterium* C58C1 harboring plasmids containing the recombinant BMV vectors yielded much stronger and persistent silencing than those induced by mechanical inoculation of the plant with recombinant BMV vector from an intermediate host or viral RNA transcripts produced as described previously (21). This system allows inoculation of virus vector without the need for expensive in vitro transcription from plasmids. In addition, further modification of this vacuum infiltration procedure may allow researchers to conduct VIGS studies in other grass species because of the efficiency of infection. For BSMV, a recombinant vector also has been constructed using a 35S promoter and inoculated using particle bombardment, again avoiding costly in vitro transcription (20). In this instance, the three BSMV genomic sequences reside in three plasmids, while for BMV, the three genomic sequences reside in two plasmids. The BSMV system was utilized to show that silencing of BLUFENSIN1 enhanced plant resistance in compatible interactions involving the causal agent of powdery mildew disease in barley (20).

We have further improved the accumulation of the DNA-based BMV silencing vector by adding the P19 gene sequence from *Tomato bushy stunt virus* (TBSV) into the plasmid encoding the viral vector. The P19 protein of TBSV is a strong suppressor of RNA silencing (78–81). Independent expression of P19 and BMV RNA 3 from the same binary vector resulted in a higher accumulation of the virus in *Agrobacterium*-infiltrated leaves, resulting in more efficient systemic infection and RNA silencing in plants (X.S. Ding and R.S. Nelson, unpublished data). The concept of utilizing a silencing suppressor to increase virus vector accumulation and associated gene silencing has been successfully utilized through stable expression of the *Tobacco mosaic virus* 126 kDa silencing suppressor protein in *Nicotiana* sp. and challenge of these plants with a TRV silencing vector (82).

Many RNA and DNA viruses have been identified and cloned from naturally infected monocotyledonous hosts (16), and some of them have been modified as transient expression vectors to express foreign genes in leaves of monocotyledons (66, 83–85). It is likely that more RNA and DNA viruses can be found that are suitable VIGS vectors for a wide range of monocotyledonous species. Any potential viral silencing vector should not induce severe disease symptoms in its host. It is also worthwhile to identify new viral vectors that are not transmitted by insect vectors and seed so that studies can be conducted under less stringent governmental regulations. For example, an infectious clone of *Panicum mosaic virus* (PMV) has been constructed (X.S. Ding and R.S. Nelson, unpublished). PMV is a single-stranded RNA virus (genus *Panicovirus*) and causes mild mosaic symptoms in many monocotyledonous species including *S. italica* and maize cultivar Oh28 (86, 87). PMV is readily transmitted among plants through mechanical inoculation but not through insect vectors and seed (86, 88). Modifications have been made to the PMV infectious clone to allow foreign gene fragment expression, and the usefulness of this virus to serve as a silencing vector is under investigation (X.S. Ding and R.S. Nelson, unpublished). Recently, a series of *Wheat streak mosaic virus*-based vectors were constructed by Tatineni and coworkers (85). Using these vectors, they were able to express the GFP gene in multiple grass species including barley, maize, and wheat (85). Because this virus encodes an RNA silencing suppressor (i.e., P1 protein: (89)), its application as a VIGS vector for grasses requires careful evaluation. This is also true for any "new" virus vector that is under construction. *Foxtail mosaic virus* (FoMV) is a member of genus *Potexvirus* and can cause infection in many species of Gramineae (90). This virus also has been modified and used as an expressing vector in several monocotyledons (91). Because both coat protein and the triple gene block ORFs were removed from the viral genome, the mutant virus expressed from this vector is defective in cell-to-cell and long-distance movement in its host plant. Consequently, this vector can only be used to express foreign genes or gene fragments in individual cells. Further modifications are needed prior to use of this vector for gene silencing in any host plant.

5. Conclusions

VIGS in grasses is a functioning system that already has yielded significant findings in gene function studies (17, 20, 25, 34, 46). However, it is not optimized in these hosts. The existing BSMV and BMV vectors are now modified for ligation-free cloning or expression behind a 35S promoter to reduce time and cost of the procedure (20, 22, 31), but no single vector has been modified for both ligation-free cloning and expression behind a plant-active promoter. These modifications are also necessary to allow higher throughput forward and reverse genetic screens. In regard to environmental considerations, when growing the plants after inoculation, there is a general consensus that lower temperatures provide better silencing phenotypes and target gene silencing (24, 47). Orientation of the gene insert also is an important consideration with the antisense orientation being the apparent no-cost optimum ((25, 57) Fig. 1).

Maintaining the plant gene insert in the virus vector appears to be the most important factor to control in order to obtain good silencing, being very closely correlated with the appearance of small RNAs, silencing of target mRNA, and maintenance of a visual silencing phenotype (24, 47). It is clear that insert stability is influenced both by fragment size and nucleotide constitution of the insert (24, 57). Lack of insert stability due to insert size likely explains the poor silencing obtained when two genes were targeted per insert as opposed to one (47). Maintenance of the insert may be influenced by both host and viral factors and certainly could explain variations between hosts in silencing efficiency. Although variation between host cultivars has been explored to identify those giving the best phenotype, modifications of viruses to enhance accumulation and insert stability are only in the early stages of study. Accumulation of virus vector has been enhanced by passage through a high-titer intermediate host (33, 34). Expressing a silencing suppressor at low levels in transgenic *N. benthamiana* also has been shown to improve silencing phenotypes, likely due to increasing the virus vector levels (82). Using this plant line as the intermediate host should further improve findings for grasses during VIGS by increasing the inoculum titer of the virus vector for the grass host. To improve insert stability, analysis of literature describing the structure of the viral RNA and *cis*- and *trans*-acting viral factors that influence virus accumulation and recombination rates is essential. Finding new viruses that maintain inserts better than current vectors do is certainly one way to improve the system for grasses. However, modifying existing and any new virus vectors at the molecular level to maintain stability will lead to the greatest advances toward optimizing this already powerful procedure for gene function studies.

Acknowledgments

The authors thank Kiran Mysore, Clemencia M. Rojas, and Muthappa Senthil-Kumar for critical review of the manuscript. This work was supported by The BioEnergy Science Center, which is a U.S. Department of Energy Bioenergy Research Center supported by the Office of Biological and Environmental Research in the DOE Office of Science, and the Samuel Roberts Noble Foundation, Inc.

Note added in proof: A BSMV vector modified for both ligation-free cloning and expression behind a plant-active promoter was recently published: Yuan et al. 2011, PLoS ONE 6, e26468.

References

1. Buell CR (2009) Poaceae genomes: going from unattainable to becoming a model clade for comparative plant genomics. Plant Physiol 149:111–116
2. Doust AN, Kellogg EA, Devos KM et al (2009) Foxtail millet: a sequence-driven grass model system. Plant Physiol 149:137–141
3. Paterson AH, Bowers JE, Bruggmann R et al (2009) The *Sorghum bicolor* genome and the diversification of grasses. Nature 457:551–556
4. Vogel JP, Garvin DF, Mockler TC et al (2010) Genome sequencing and analysis of the model grass *Brachypodium distachyon*. Nature 463:763–768
5. Burch-Smith TM, Anderson JC, Martin GB et al (2004) Applications and advantages of virus-induced gene silencing for gene function studies in plants. Plant J 39:734–746
6. Tadege M, Ratet P, Mysore KS (2005) Insertional mutagenesis: a Swiss Army knife for functional genomics of *Medicago truncatula*. Trends Plant Sci 10:229–235
7. Kondou Y, Higuchi M, Matsui M (2010) High-throughput characterization of plant gene functions by using gain-of-function technology. Annu Rev Plant Biol 61:373–393
8. Andrea K, Rownak A (2003) Physical and chemical mutagenesis. Methods Mol Biol 236:189–203
9. Gilchrist EJ, Haughn GW (2005) TILLING without a plough: a new method with applications for reverse genetics. Curr Opin Plant Biol 8:211–215
10. Parinov S, Sundaresan V (2000) Functional genomics in Arabidopsis: large-scale insertional mutagenesis complements the genome sequencing project. Curr Opin Biotechnol 11:157–161
11. Senthil-Kumar M, Anand A, Uppalapati SR et al (2008) Virus-induced gene silencing and its applications. CAB Rev Perspect Agric Vet Sci Nutr Nat Resour 3:11
12. Becker A, Lange M (2010) VIGS genomics goes functional. Trends Plant Sci 15:1–4
13. Cakir C, Gillespie ME, Scofield SR (2010) Rapid determination of gene function by virus-induced gene silencing in wheat and barley. Crop Sci 50:S77–S84
14. Robertson D (2004) VIGS vectors for gene silencing: many targets, many tools. Annu Rev Plant Biol 55:495–519
15. Baulcombe DC (1999) Fast forward genetics based on virus-induced gene silencing. Curr Opin Plant Biol 2:109–113
16. Scofield SR, Nelson RS (2009) Resources for virus-induced gene silencing in the grasses. Plant Physiol 149:152–157
17. Held MA, Penning B, Brandt AS et al (2008) Small-interfering RNAs from natural antisense transcripts derived from a cellulose synthase gene modulate cell wall biosynthesis in barley. Proc Natl Acad Sci USA 105:20534–20539
18. Holzberg S, Brosio P, Gross C et al (2002) *Barley stripe mosaic virus*-induced gene silencing in a monocot plant. Plant J 30:315–327
19. Tai YS, Bragg J, Edwards MC (2005) Virus vector for gene silencing in wheat. Biotechniques 39:310–312
20. Meng Y, Moscou MJ, Wise RP (2009) Blufensin1 negatively impacts basal defense in response to barley powdery mildew. Plant Physiol 149:271–285
21. Ding XS, Schneider WL, Chaluvadi SR et al (2006) Characterization of a *Brome mosaic virus* strain and its use as a vector for gene silencing in monocotyledonous hosts. Mol Plant Microbe Interact 19:1229–1239
22. Pacak A, Strozycki PM, Barciszewska-Pacak M et al (2010) The *Brome mosaic virus*-based recombination vector triggers a limited gene silencing response depending on the orientation of the inserted sequence. Arch Virol 155:169–179
23. Purkayastha A, Mathur S, Verma V et al (2010) Virus-induced gene silencing in rice using a vector derived from a DNA virus. Planta 232:1531–1540
24. Bruun-Rasmussen M, Madsen CT, Jessing S et al (2007) Stability of *Barley stripe mosaic virus*-induced gene silencing in barley. Mol Plant Microbe Interact 20:1323–1331
25. Scofield SR, Amanda LH, Brandt S et al (2005) Development of a virus-induced gene silencing system for hexaploid wheat and its use in functional analysis of the Lr21-mediated leaf rust resistance pathway. Plant Physiol 138:2165–2173

26. Renner T, Bragg J, Driscoll HE et al (2009) Virus-induced gene silencing in the culinary ginger (*Zingiber officinale*): an effective mechanism for down-regulating gene expression in tropical monocots. Mol Plant 2:1084–1094
27. Wang X, Cao A, Yu C et al (2010) Establishment of an effective virus-induced gene silencing system with BSMV in *Haynaldia villosa*. Mol Biol Rep 37:967–972
28. Demircan T, Akkaya M (2010) Virus-induced gene silencing in *Brachypodium distachyon*, a model organism for cereals. Plant Cell Tissue Organ Cult 100:91–96
29. Pacak A, Geisler K, Jorgensen B et al (2010) Investigations of *barley stripe mosaic virus* as a gene silencing vector in barley roots and in *Brachypodium distachyon* and oat. Plant Methods 6:26
30. Hema M, Kao CC (2004) Template sequence near the initiation nucleotide can modulate *Brome mosaic virus* RNA accumulation in plant protoplasts. J Virol 78:1169–1180
31. Ding XS, Ballard K, Nelson RS (2009) Improving virus-induced gene silencing (VIGS) in rice through *Agrobacterium* infiltration. In: Proceedings of the 14th International Congress on Molecular Plant-Microbe Interactions 7, 19–23 July, Quebec City, Quebec, CA
32. Alejska M, Figlerowicz M, Malinowska N et al (2005) A universal BMV-based RNA recombination system-how to search for general rules in RNA recombination. Nucleic Acids Res 33:e105
33. Ding XS, Rao CS, Nelson RS (2007) Analysis of gene function in rice through virus-induced gene silencing. Methods Mol Biol 354:145–160
34. Van der Linde K, Kastner C, Kumlehn J et al (2011) Systemic virus-induced gene silencing allows functional characterization of maize genes during biotrophic interaction with *Ustilago maydis*. New Phytol 189:471–483
35. Loutre C, Wicker T, Travella S et al (2009) Two different CC-NBS-LRR genes are required for Lr10-mediated leaf rust resistance in tetraploid and hexaploid wheat. Plant J 60:1043–1054
36. Zhou HB, Li SF, Deng ZY et al (2007) Molecular analysis of three new receptor-like kinase genes from hexaploid wheat and evidence for their participation in the wheat hypersensitive response to stripe rust fungus infection. Plant J 52:420–434
37. Zhang L, Lavery L, Gill U et al (2009) A cation/proton-exchanging protein is a candidate for the barley NecS1 gene controlling necrosis and enhanced defense response to stem rust. Theor Appl Genet 118:385–397
38. Hein I, Pacak MB, Hrubikova K et al (2005) Virus-induced gene silencing-based functional characterization of genes associated with powdery mildew resistance in barley. Plant Physiol 138:2155–2164
39. Shen QH, Saijo Y, Mauch S et al (2007) Nuclear activity of MLA immune receptors links isolate-specific and basal disease-resistance responses. Science 315:1098–1103
40. Cao AH, Xing LP, Wang XY et al (2011) Serine/threonine kinase gene Stpk-V, a key member of powdery mildew resistance gene Pm21, confers powdery mildew resistance in wheat. Proc Natl Acad Sci USA 108: 7727–7732
41. Manning VA, Chu AL, Scofield SR et al (2010) Intracellular expression of a host-selective toxin, ToxA, in diverse plants phenocopies silencing of a ToxA-interacting protein, ToxABP1. New Phytol 187:1034–1047
42. Wang G-F, Wei X, Fan R et al (2011) Molecular analysis of common wheat genes encoding three types of cytosolic heat shock protein 90 (Hsp90): functional involvement of cytosolic Hsp90s in the control of wheat seedling growth and disease resistance. New Phytol 191:418–431
43. Delventhal R, Zellerhoff N, Schaffrath U (2011) Barley stripe mosaic virus-induced gene silencing (BSMV-IGS) as a tool for functional analysis of barley genes potentially involved in nonhost resistance. Plant Signal Behav 6:867–869
44. Van Eck L, Schultz T, Leach JE et al (2010) Virus-induced gene silencing of WRKY53 and an inducible phenylalanine ammonia-lyase in wheat reduces aphid resistance. Plant Biotechnol J 8:1023–1032
45. Oikawa A, Rahman A, Yamashita T et al (2007) Virus-induced gene silencing of P23k in barley leaf reveals morphological changes involved in secondary wall formation. J Exp Bot 58:2617–2625
46. Shi Y, Qin Y, Cao Y et al (2011) Influence of an m-type thioredoxin in maize on potyviral infection. Eur J Plant Pathol 131:1–10
47. Cakir C, Tor M (2010) Factors influencing *Barley stripe mosaic virus*-mediated gene silencing in wheat. Physiol Mol Plant Pathol 74:246–253
48. Tufan HA, Stefanato FL, McGrann GRD et al (2011) The *Barley stripe mosaic virus* system used for virus-induced gene silencing in cereals differentially affects susceptibility to fungal pathogens in wheat. J Plant Physiol 168:990–994
49. Lane LC (1977) *Brome mosaic virus*. CMI/AAB descriptions of plant viruses no. 180

50. Moreno IM, Garcia-Arenal F (2004) Genus Bromovirus, In: Lapierre PA and Signoret P.-A. (ed.) Viruses and virus diseases of Poaceae (Gramineae). INRA editions, Paris, pp 352–355
51. Chellappan P, Vanitharani R, Ogbe F et al (2005) Effect of temperature on geminivirus-induced RNA silencing in plants. Plant Physiol 138:1828–1841
52. Fu DQ, Zhu BZ, Zhu HL et al (2006) Enhancement of virus-induced gene silencing in tomato by low temperature and low humidity. Mol Cells 21:153–160
53. Igarashi A, Yamagata K, Sugai T et al (2009) *Apple latent spherical virus* vectors for reliable and effective virus-induced gene silencing among a broad range of plants including tobacco, tomato, *Arabidopsis thaliana*, cucurbits, and legumes. Virology 386:407–416
54. Szittya G, Silhavy D, Molnar A et al (2003) Low temperature inhibits RNA silencing-mediated defence by the control of siRNA generation. EMBO J 22:633–640
55. Tuttle JR, Idris AM, Brown JK et al (2008) Geminivirus-mediated gene silencing from *Cotton leaf crumple virus* is enhanced by low temperature in cotton. Plant Physiol 148:41–50
56. Ding XS, Flasinski S, Nelson RS (1999) Infection of barley by *Brome mosaic virus* is restricted predominantly to cells in and associated with veins through a temperature-dependent mechanism. Mol Plant Microbe Interact 12:615–623
57. Lacomme C, Hrubikova K, Hein I (2003) Enhancement of virus-induced gene silencing through viral-based production of inverted-repeats. Plant J 34:543–553
58. Senthil-Kumar M, Mysore KS (2011) Virus-induced gene silencing can persist for more than 2 years and also be transmitted to progeny seedlings in *Nicotiana benthamiana* and tomato. Plant Biotechnol J 9:797–806
59. Yamagishi N, Yoshikawa N (2009) Virus-induced gene silencing in soybean seeds and the emergence stage of soybean plants with *Apple latent spherical virus* vectors. Plant Mol Biol 71:15–24
60. Kim MJ, Kao C (2001) Factors regulating template switch in vitro by viral RNA-dependent RNA polymerases: implications for RNA-RNA recombination. Proc Natl Acad Sci USA 98:4972–4977
61. Lai MMC (1992) RNA recombination in animal and plant viruses. Microbiol Rev 56:61–79
62. Shapka N, Nagy PD (2004) The AU-rich RNA recombination hot spot sequence of *brome mosaic virus* is functional in tombusviruses: implications for the mechanism of RNA recombination. J Virol 78:2288–2300
63. Shi BJ, Symons RH, Palukaitis P (2008) The cucumovirus 2b gene drives selection of inter-viral recombinants affecting the crossover site, the acceptor RNA and the rate of selection. Nucleic Acids Res 36:1057–1071
64. Sztuba-Solinska J, Dzianott A, Bujarski JJ (2011) Recombination of 5 subgenomic RNA3a with genomic RNA3 of *Brome mosaic bromovirus in vitro* and *in vivo*. Virology 410:129–141
65. Constantin GD, Krath BN, MacFarlane SA et al (2004) Virus-induced gene silencing as a tool for functional genomics in a legume species. Plant J 40:622–631
66. Liu Y, Schiff M, Marathe R et al (2002) Tobacco Rar1, EDS1 and NPR1/NIM1 like genes are required for N-mediated resistance to *Tobacco mosaic virus*. Plant J 30:415–429
67. Lu R, Martin-Hernandez AM, Peart JR et al (2003) Virus-induced gene silencing in plants. Methods 30:296–303
68. Wang CC, Cai XZ, Wang XM et al (2006) Optimisation of *Tobacco rattle virus*-induced gene silencing in Arabidopsis. Funct Plant Biol 33:347–355
69. Orzaez D, Medina A, Torre S et al (2009) A visual reporter system for virus-induced gene silencing in tomato fruit based on anthocyanin accumulation. Plant Physiol 150:1122–1134
70. Ryu CM, Anand A, Kang L et al (2004) Agrodrench: a novel and effective agroinoculation method for virus-induced gene silencing in roots and diverse solanaceous species. Plant J 40:322–331
71. Wroblewski T, Tomczak A, Michelmore R (2005) Optimization of *Agrobacterium*-mediated transient assays of gene expression in lettuce, tomato and Arabidopsis. Plant Biotechnol J 3:259–273
72. Xie YH, Zhu BZ, Yang XL et al (2006) Delay of postharvest ripening and senescence of tomato fruit through virus-induced LeACS2 gene silencing. Postharvest Biol Technol 42:8–15
73. Dasgupta I, Hull R, Eastop S et al (1991) *Rice tungro bacilliform virus*-DNA independently infects rice after *Agrobacterium*-mediated transfer. J Gen Virol 72:1215–1221
74. Grimsley N, Thomas H, Davies JW et al (1987) *Agrobacterium*-mediated delivery of infectious *Maize streak virus* into maize plants. Nature 325:177–179
75. Hayes RJ, Macdonald H, Coutts RHA et al (1988) Agroinfection of Triticum aestivum with cloned DNA of Wheat dwarf virus. J Gen Virol 69:891–896

76. Martin DP, Rybicki EP (2000) Improved efficiency of *Zea mays* agroinoculation with *Maize streak virus*. Plant Dis 84:1096–1098
77. Shen W-H, Hohn B (1994) Amplification and expression of the β-glucuronidase gene in maize plants by vectors based on *Maize streak virus*. Plant J 5:227–236
78. Dunoyer P, Lecellier CH, Parizotto EA et al (2004) Probing the microRNA and small interfering RNA pathways with virus-encoded suppressors of RNA silencing. Plant Cell 16:1235–1250
79. Lakatos L, Szittya G, Silhavy D et al (2004) Molecular mechanism of RNA silencing suppression mediated by p19 protein of *Tombusviruses*. EMBO J 23:876–884
80. Qiu WP, Park JW, Scholthof HB (2002) *Tombusvirus* p19-mediated suppression of virus-induced gene silencing is controlled by genetic and dosage features that influence pathogenicity. Mol Plant Microbe Interact 15:269–280
81. Ye K, Malinina L, Patel DJ (2003) Recognition of small interfering RNA by a viral suppressor of RNA silencing. Nature 426:874–878
82. Harries PA, Palanichelvam K, Bhat S et al (2008) *Tobacco mosaic virus* 126-kDa protein increases the susceptibility of *Nicotiana tabacum* to other viruses and its dosage affects virus-induced gene silencing. Mol Plant Microbe Interact 21:1539–1548
83. Choi I-R, Stenger DC, Morris TJ et al (2000) A plant virus vector for systemic expression of foreign genes in cereals. Plant J 23:547–555
84. Haupt S, Duncan GH, Holzberg S et al (2001) Evidence for symplastic phloem unloading in sink leaves of barley. Plant Physiol 125:209–218
85. Tatineni S, McMechan AJ, Hein GL et al (2011) Efficient and stable expression of GFP through *Wheat streak mosaic virus*-based vectors in cereal hosts using a range of cleavage sites: formation of dense fluorescent aggregates for sensitive virus tracking. Virology 410:268–281
86. Niblett CL, Paulsen AQ (1975) Purification and further characterization of *Panicum Mosaic Virus*. Phytopathology 65:1157–1160
87. Sill JWH, Talens LT (1962) New hosts and characteristics of the *Panicum mosaic virus*. Plant Dis Rep 46:780–783
88. Niblett CL, Paulsen AQ, Toler RW (1977) *Panicum mosaic virus*. CMI/AAB description of plant viruses no. 177
89. Stenger DC, Young BA, Qu F et al (2007) *Wheat streak mosaic virus* P1, not HC-Pro, facilitates disease synergism and suppression of post-transcriptional gene silencing. Phytopathology 97:S111
90. Short MN (1983) *Foxtail mosaic virus*. CMI/AAB description of plant viruses. no. 264
91. Liu Z, Kearney C (2010) An efficient *Foxtail mosaic virus* vector system with reduced environmental risk. BMC Biotechnol 10:88

Chapter 3

Virus-Induced Gene Silencing for Rice Using Agroinoculation

Arunima Purkayastha, Shweta Sharma, and Indranil Dasgupta

Abstract

Virus-induced gene silencing (VIGS) is a reverse genetics technique that is based on the RNA-mediated defense against viruses in plants. VIGS is a method of gene knockdown triggered by a replicating viral nucleic acid engineered to carry a host gene to be silenced. While there are a number of excellent VIGS vectors available for dicots, only a few are available for monocots. Here, we describe the detailed method of the use of a newly developed VIGS vector for rice, based on the rice-infecting *Rice tungro bacilliform virus*, a pararetrovirus with dsDNA genome. Using a method based on *Agrobacterium*-mediated injection of the VIGS construct at the meristematic region of young rice plants, silencing of target genes can be achieved and the silenced phenotype can be visualized in 3 weeks.

Key words: VIGS, Rice, *Rice tungro bacilliform virus*, Gene silencing, Agroinoculation

1. Introduction

RNA silencing is a sequence-specific degradation of mRNA, induced by double-stranded RNA (dsRNA) that leads to the silencing of gene expression (1). In plants, RNA silencing is commonly seen during virus infection and is believed to be an inherent defense response against invading viruses. Virus-induced gene silencing (VIGS), based on the RNA silencing mediated defense response against viruses, is a method of transient gene silencing in plants using viral vectors. During the course of infection, viruses, irrespective of containing RNA or DNA genomes, produce double-stranded RNA, either as replication intermediates of RNA viruses or highly structured regions of viral transcripts, both of which are efficient triggers of gene silencing (2, 3). In plants, recombinant viruses carrying a complete or part of an endogenous plant gene, as a cDNA fragment, in addition to getting targeted, trigger the silencing of homologous host transcripts by RNA silencing. The above phenomenon creates a transient "knockdown" of the gene,

Annette Becker (ed.), *Virus-Induced Gene Silencing: Methods and Protocols*, Methods in Molecular Biology, vol. 975,
DOI 10.1007/978-1-62703-278-0_3, © Springer Science+Business Media New York 2013

enabling reverse genetic analysis of the targeted plant genes to be performed (4).

Although VIGS has been used extensively for dicots, examples of the use of VIGS vectors in monocots are limited. To date, three viruses have been modified to function as VIGS vectors in monocots: a vector based on *Barley stripe mosaic virus* for barley, wheat, and maize (5); a vector based on *Brome mosaic virus* for rice, in addition to barley and maize (6); and, more recently, one based on *Rice tungro bacilliform virus* (RTBV) for rice (7). The first two vectors are based on RNA viruses, and their inoculations involve the generation of in vitro transcripts, followed by inoculation to plants. The RTBV-based vector is based on a DNA virus, and inoculation involves agroinoculation procedures.

Rice is by far the world's most important cereal crop and is particularly important in the South and Southeast Asian countries. The entire genome of rice has been sequenced, and research is now focused on annotating over 30,000 predicted genes (8). A VIGS-based gene silencing system for rice, which allows high throughput genome-wide gene silencing, is highly desirable. The RTBV-based VIGS vector fits into this category. Here, we describe the detailed protocol for use of the RTBV-based VIGS vector for gene silencing in rice.

2. Materials

The plant gene to be targeted by the RTBV-based VIGS vector (pRTBV-MVIGS) needs to be cloned into the multiple cloning sites (MCS) in antisense orientation. The construct with the cloned insert is then transformed into *Agrobacterium tumefaciens* strain EHA105 by chemical transformation method. The transformant *Agrobacterium* colonies obtained are confirmed for the presence of the insert by colony PCR. An *Agrobacterium* colony with the desired insert is grown in LB broth with the appropriate antibiotic selection. The cells are harvested by centrifugation and then resuspended into a buffer. This suspension is used for agroinoculation of 15-day-old rice plants germinated and grown in Yoshida's medium. The appearance of phenotype is scored for 21–35 days postinoculation (dpi). The downregulation of the target gene can be validated by performing real-time PCR using gene-specific primers.

2.1. Plant Growth Conditions

1. Yoshida's medium (9) (see Note 1).
2. Test tubes.
3. Culture room/chamber maintained at 27°C and a 16 h light and 8 h dark cycle using artificial lighting with 70% humidity.

2.2. Cloning of Target Gene into the VIGS Vector

1. Sterilized microcentrifuge tubes, tips, etc.
2. Electrophoresis equipment and reagents.
3. Restriction enzymes and buffers, water baths.
4. VIGS vector pRTBV-MVIGS (7).
5. Commercially available total plant RNA isolation reagent (TRIzol Reagent, Invitrogen, Carlsbad, CA, USA), MOPS buffer (see page 7/Subheading 3.1.1) (see Notes 2–4).
6. Commercially available cDNA synthesis kit (High capacity cDNA reverse transcription kit, Applied Biosystems, Carlsbad, CA, USA).

   Nanodrop (NanoVue Spectrophotometer V1.7.3, GE Healthcare, England).
7. High-fidelity Phusion polymerase (Finnzymes, Espoo, Finland) and buffers, dATP and dNTP, gene-specific primers, thermal cyclers.
8. HiYield Gel extraction/PCR purification kit (Real Biotech Corporation (RBC), Taipei County, Taiwan).
9. InsT/A Cloning kit (Fermentas, Ontario, Canada).
10. T4 DNA ligase and buffer (Fermentas).
11. Transformation-competent *Escherichia coli* (DH5α) cells.
12. Reagents for transformation and selection of recombinant clones, Luria-Bertani (LB) broth and agar, Petri dishes, appropriate antibiotic stocks, sterilized toothpicks, water baths, orbital shakers, centrifuge machines, spreader, incubator, culture vials, restriction enzymes, and buffers.
13. Reagents for plasmid isolation (10).

2.3. Mobilization of the Recombinant VIGS Vector into A. tumefaciens by Chemical Transformation

1. Sterilized microcentrifuge tubes, tips, etc.
2. Transformation-competent Agrobacterium (EHA105) cells (11).
3. Reagents for transformation: Luria-Bertani (LB) broth and agar, Petri dishes, antibiotic stocks, sterilized toothpicks, water baths, orbital shakers, centrifuge machines, spreader, incubator, culture vials, restriction enzymes, and buffers.
4. For colony PCR screening of recombinant clones: Taq DNA polymerase (NEB, Ipswich, MA), gene-specific primers, and thermal cycler.

2.4. Preparation of Resuspension for Agroinoculation of Plants

1. Reagents for growing primary and secondary culture of recombinant VIGS vector: LB broth, antibiotics.
2. Sterilized oak ridge tubes (Sorvall SS34 tubes).
3. Superspeed centrifuge.
4. Reagents for resuspension buffer: 10 mM $MgCl_2$ stock, 10 mM MES stock, 200 μM acetosyringone stock (see Notes 5 and 6).

2.5. Agroinoculation of Plants

1. Sterilized 1 ml syringe (DISPO VAN®, Hindustan Syringes and Medical Devices Ltd., India).
2. Whatman No. 1 filter paper (Whatman International Ltd., England).
3. Yoshida's medium (9).

2.6. Analysis of VIGS in Agroinoculated Plants by Scoring of Phenotype and Real-Time RT-PCR

1. Reagents for total RNA isolation.
2. Commercially available total plant RNA isolation reagent (TRIzol Reagent, Invitrogen).
3. Commercially available first strand cDNA synthesis kit (High capacity cDNA reverse transcription kit, Applied Biosystems).
4. Nanodrop (NanoVue Spectrophotometer, V1.7.3, GE Healthcare, England).
5. Gene-specific primers that are designed to exclude the sequence cloned in the VIGS vector and primers for internal control.
6. Commercially available real-time PCR kit (SYBR Green PCR Master mix, Applied Biosystems).
7. MicroAmp® Optical 96-well reaction plate (Applied Biosystems).
8. MicroAmp® Optical adhesive film kit (Applied Biosystems).
9. Real-time PCR cycler (ABI Prism® 7000 Sequence Detection System, Applied Biosystems).

3. Methods

3.1. Cloning of Target Gene into the VIGS Vector

A partial cDNA corresponding to the coding region of the target gene is cloned in antisense orientation in the multiple cloning sites of the VIGS vector.

3.1.1. Amplification and Cloning of the Target Gene into VIGS Vector

1. Precool mortar and pestle by addition of liquid nitrogen. Grind fresh leaf material frozen in liquid nitrogen to a fine powder.
2. Isolate total RNA using TRIzol Reagent (Invitrogen) as per manufacturer's instructions.
3. Quantify the RNA samples by measuring the absorbance of the sample at 260 nm using a Nanodrop. Calculate the amount of RNA according to the correlation: Absorbance of 1.0 at OD_{260} equivalent to 40 μg/ml of RNA.
4. Perform a MOPS gel electrophoresis for estimating the quality of the RNA preparations. Aliquot sample volume equivalent to 1 or 2 mg of RNA and mix with 3 volumes of premix solution (1× MOPS {3-(*N*-morpholino)-propanesulfonic acid}: formaldehyde: formamide = 1:3.5:10 and 250 μg/ml ethidium bromide)

(10× MOPS = 400 mM MOPS, 99.6 mM sodium acetate and 20 mM EDTA, pH 7.0) and 1/10 volume of sample loading buffer (0.25% bromophenol blue, 0.25% xylene cyanol, 1 mM EDTA, and 50% glycerol).

5. Denature the mixture at 65°C for 10 min and then snap cool in ice.
6. Resolve the sample by electrophoresis in a 1.2% agarose gel containing 1.1% formaldehyde, prepared in 1× MOPS buffer. Perform the electrophoresis in 1× MOPS buffer at constant voltage (100 V) for 1 h. Visualize the gel on a UV transilluminator.
7. Amplify the total cDNA using High capacity cDNA reverse transcription kit (Applied Biosystems) as per manufacturer's instructions.
8. Use gene-specific primers to amplify a partial cDNA of the target gene. The primers should be designed to include the site PacI and MluI at the 5¢ end of the forward and reverse primer, respectively, for cloning into pRTBV-MVIGS.
9. Set up a PCR with Phusion polymerase (Finnzymes) according to manufacturer's instructions.
10. Mix contents well and centrifuge the tube briefly to collect the contents at the bottom.
11. Set up the following program on a thermal cycler: 98°C, 3 min/30× (98°C, 10 s/55°C, 30 s/72°C, 10 s)/72°C, 7 min.
12. After the completion of the program, keep the reaction at 4°C.
13. Electrophorese a 5 ml aliquot of the PCR product in 1% agarose gel (1× TBE, 0.5 mg/ml of ethidium bromide) in parallel with a DNA molecular weight standard (see Note 11). Visualize the DNA in a gel documentation system to confirm the amplification of the correct sized product.
14. Purify the remaining PCR product to remove the proofreading enzyme using a PCR purification kit (RBC) as per the manufacturer's instruction. Elute the purified product in 30 ml of autoclaved MQ water.
15. Check a 2 ml aliquot of the purified PCR product on 1% agarose gel (1× TBE, 0.5 μg/ml of ethidium bromide) in parallel with a DNA molecular weight standard. Visualize the DNA in a gel documentation system to estimate the quantity of the purified PCR product.
16. Set up the following reaction for A-tailing: 28 ml purified PCR product, 1.5 μl dATP (10 mM stock), 1–2 units of Taq DNA polymerase enzyme, 5 μl 10× Taq polymerase buffer and make the volume to 50 μl with water. Incubate the mix at 72°C for 30 min. Snap cool reaction in ice and store in −20°C before ligation with T/A vector.

17. The A-tailed PCR product can be cloned into vectors having T overhangs using InsT/A cloning kit (Fermentas) as per manufacturer's instructions. Plate on LB plates containing ampicillin, IPTG (isopropylthiogalactoside), and X-gal (5-Bromo-4-chloro-3-indolyl β-d-galactoside) (see Notes 7–9).
18. Pick 10 white colonies and inoculate 2 ml LB medium containing 50 mg/ml ampicillin and incubate the bacterial cultures at 37°C in an orbital shaker (200 rpm) overnight (see Note 16).
19. Isolate plasmid from the bacterial cultures by plasmid mini preparations using alkaline lysis method (12).
20. Digest the plasmid preparations with restriction enzymes MluI and PacI and check the digested product on 1% agarose gel (1× TBE, 0.5 μg/ml of ethidium bromide) in parallel with a DNA molecular weight standard and the PCR product as a positive control. Select clones that show the release of the correct sized insert (see Note 10).
21. Digest sufficient amount of the recombinant T/A clone with MluI and PacI and check the digested product on a fresh 1% agarose gel (1× TBE, 0.5 μg/ml of ethidium bromide) in parallel with a DNA molecular weight standard. Elute the desired fragment from the gel using gel extraction kit (RBC) as per the manufacturer's instructions.
22. Digest 5 μg of the VIGS vector (pRTBV-MVIGS) with MluI and PacI. Subject to electrophoresis and recover the double-digested fragment using commercially available gel extraction kit (RBC).
23. Assess the approximate concentrations of the vector and the insert DNA by gel electrophoresis using an aliquot of each preparation along with a DNA size marker of known concentration.
24. Set up ligation reaction using the vector and the insert fragment in 3:1 ratio.
25. Transform the ligation mix from previous step in competent *E. coli* DH5a cells and plate on LB plates supplemented with appropriate antibiotics (kanamycin at 50 μg/ml).
26. For screening of transformants, inoculate in 2 ml LB medium containing 50 μg/ml kanamycin and incubate the bacterial cultures at 37°C in an orbital shaker (200 rpm) overnight.
27. Isolate plasmid from the bacterial cultures by plasmid mini preparations using alkaline lysis method.
28. Digest the plasmid preparations with MluI and PacI and check the digested product on 1% agarose gel (1× TBE, 0.5 μg/ml of ethidium bromide) in parallel with a DNA molecular weight standard and the PCR product as a positive control. Select clones that show the release of the correct size insert.

3.2. Mobilization of the Recombinant VIGS Vector into A. tumefaciens by Chemical Transformation

The recombinant VIGS vector containing the partial cDNA of the target gene in reverse orientation is transformed into *Agrobacterium* strain EHA105, and the transformants are screened by colony PCR method.

3.2.1. Preparation of Competent Agrobacterium Cells

1. Inoculate 5 ml of LB broth containing 50 μg/ml of rifampicin with *Agrobacterium* strain EH105 (see Note 13). Incubate with shaking in dark at 28°C for 48 h.
2. Start a secondary culture by transferring 1 ml of the primary culture to 50 ml LB medium containing 50 μg/ml of rifampicin in a 250 ml conical flask. Incubate at 28°C in dark with shaking till the $O.D_{600}$ of the culture reaches 1.0.
3. Transfer the culture to sterilized centrifuge tubes (Sorvall SS34 tubes) and pellet the cells by centrifuging at 5,000 × g for 10 min in an appropriate rotor at 4°C. Discard the supernatant.
4. Gently resuspend the cells in 1 ml of filter sterilized, ice-cold 100 mM $CaCl_2$. Repeat centrifuge as described in the previous step. Discard the supernatant.
5. Resuspend the pellet gently in 425 ml of ice-cold 100 mM $CaCl_2$ and 75 μl of sterilized 100% glycerol.
6. Aliquot 50 μl of the final resuspension into prechilled sterilized microcentrifuge tubes.
7. Freeze the aliquots in liquid nitrogen and store competent cells at –70°C.

3.2.2. Transformation of Competent Agrobacterium Cells

1. Add 1–2 μg purified plasmid DNA dissolved in sterile MQ water to a vial of transformation-competent *Agrobacterium* cells thawed in ice.
2. Freeze by lowering the MCT into liquid nitrogen for 2 min.
3. Incubate the cells at 37°C for 5 min.
4. Incubate the cells in ice for 10 min.
5. Add 1 ml of LB medium to the tube. Divide the contents into two sterile MCT's and incubate with 200 rpm shaking at 28°C for overnight.
6. Centrifuge the tubes at 5,000 × *g* for 5 min to pellet the cells.
7. Gently resuspend the cells in 100 μl of LB medium and plate on LB plates supplemented with 50 μg/ml rifampicin and 50 μg/ml kanamycin. Incubate the plates at 28°C for 2–3 days until the colonies are sufficiently large.
8. For screening the transformants, perform colony PCR with target gene-specific primers.
9. Prepare a glycerol stock of the correct transformant and store at –70°C for further use.

3.3. Preparation of Resuspension for Agroinoculation of Plants

1. Using a sterile inoculation loop, streak *Agrobacterium* culture (VIGS vector containing the target gene) on LB agar plates containing 50 μg/ml rifampicin and 50 μg/ml kanamycin so as to obtain single colonies. Incubate the plates at 28°C in dark for 2–3 days until the colonies are sufficiently large.
2. Inoculate 5 ml of LB medium containing 50 μg/ml rifampicin and 50 μg/ml kanamycin. Incubate with 200 rpm shaking at 28°C in dark for 48 h.
3. Start a secondary culture by transferring 1 ml of the primary culture to 100 ml LB medium containing 50 μg/ml rifampicin and 50 μg/ml kanamycin. Incubate with 200 rpm shaking at 28°C in dark till $O.D_{600}$ reaches 0.6–0.8.
4. Transfer the culture to sterilized centrifuge tubes (Sorvall SS34 tubes) and pellet the cells by centrifuging at 5,000 × *g* for 10 min in an appropriate rotor at 4°C. Discard the supernatant.
5. Resuspend the cells in a 1/20 volume of the original culture of the resuspension buffer (see Subheading 2.4) and incubate at room temperature for 2–3 h.
6. Transfer the contents to sterilized MCT and use for agroinoculation.

3.4. Agroinoculation of Plants

3.4.1. Preparation of Plants

1. Surface sterilize the seeds by washing in 70% ethanol for 45 s, followed by 5–6 washes with copious amount of sterile double-distilled water (SDDW).
2. Spread the seeds on a fine muslin cloth held over a reservoir of Yoshida's medium such that the liquid medium just wets the muslin cloth. Grow the plants at 27°C and at 16 h light and 8 h dark cycle using artificial lighting with 70% humidity.
3. Keep adding Yoshida's medium to the reservoir to maintain the level.
4. 10–12 days post-germination, transfer individual plants to test tubes containing Yoshida's medium.
5. Use 15–18-day-old plants for agroinoculation.

3.4.2. Agroinoculation Procedure

1. Use a clinical syringe to inoculate about 50 μl of the bacterial suspension at the meristematic region located at the crown region of the plant (Fig. 1a–c). Take care to insert the needle vertically down the stem toward the base of the plant. Ideally, the suspension should force out from near the base of the first leaf (see Notes 11–13).
2. Transfer the plants to a sterile Whatman No. 1 filter paper saturated in Yoshida's medium and placed on a solid support with its ends dipped into a reservoir containing the medium (Fig. 2) (see Note 14).
3. Cover the plants with moist tissue paper to prevent loss of water (Fig. 3).

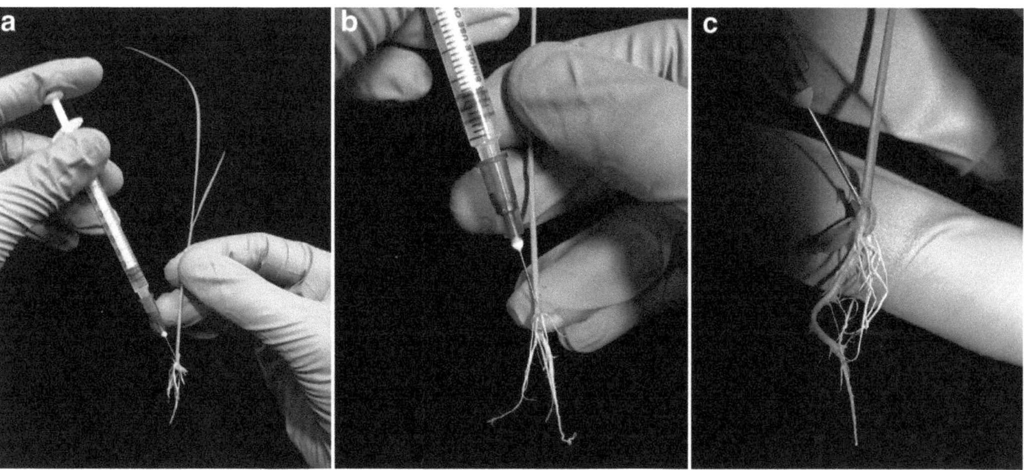

Fig. 1. Pictures showing the agroinoculation procedure of VIGS vector into the rice plants. A clinical syringe is used to inoculate the bacterial suspension at the meristematic region located at the crown region of the plant.

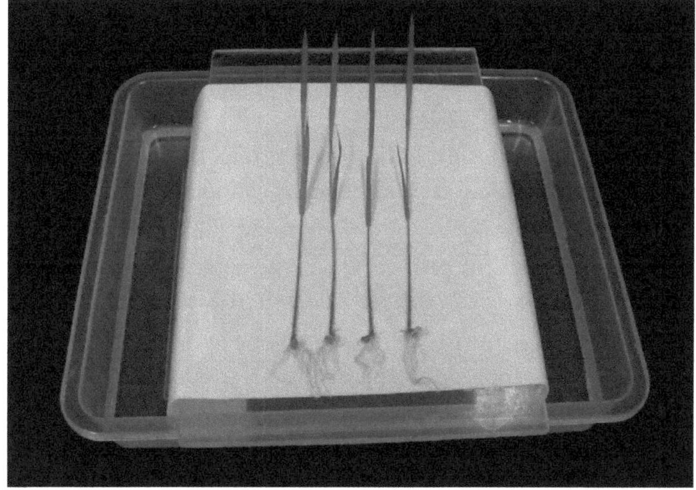

Fig. 2. Picture showing the setup used for the incubation of rice plants after agroinoculation with the VIGS vector. The setup includes a sterile Whatman No. 1 filter paper saturated in Yoshida's medium and placed on a solid support with its ends dipped into a reservoir containing the medium.

4. Maintain the agroinoculated plants at 27°C and 70% humidity for 24 h after agroinoculation (see Note 15).
5. Transfer plants to individual test tubes containing Yoshida's medium and grow at conditions mentioned earlier (see Notes 16–17).

3.5. Analysis of VIGS in Plants by Phenotype Scoring and Real-Time RT-PCR

3.5.1. Scoring the Agroinoculated Plants for the Silencing Phenotype

1. Maintain the agroinoculated plants at the conditions mentioned above in order to score for the phenotype. We routinely maintain rice plants till 10 weeks postinoculation (wpi).
2. For silencing of *phytoene desaturase*, phenotype (white streaks in emerging leaves) starts appearing in the agroinoculated plants 3–5 wpi and is maintained till 10 wpi.

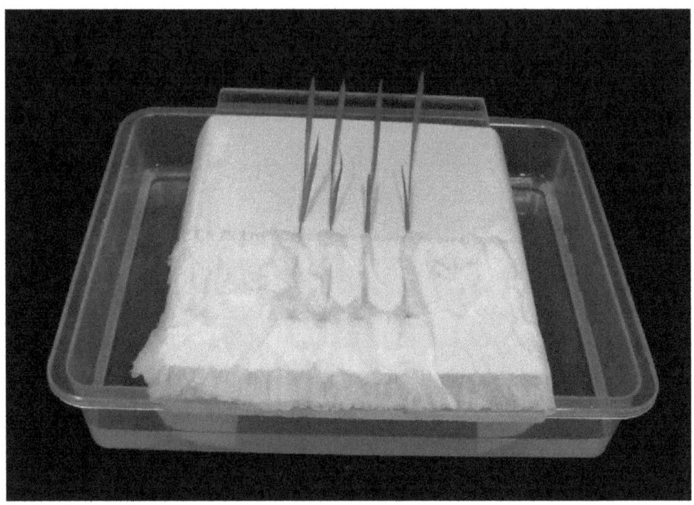

Fig. 3. Picture showing the agroinoculated rice plants covered with a moist tissue paper to prevent loss of water.

3.5.2. Validation of the Target Gene Silencing by Real-Time RT-PCR

1. Isolate total RNA from the leaf tissues showing the desired phenotype using TRIzol Reagent (Invitrogen) as per manufacturer's instructions.
2. To design PCR primers, select a unique region from the cDNA sequence of interest by using BLAST search engine in NCBI database.
3. Design primers from the selected unique region using Primer Express® software (Applied Biosystems).
4. Quantify the RNA samples by measuring the absorbance of the sample at 260 nm using a Nanodrop (NanoVue, GE Healthcare). Calculate the amount of RNA according to the correlation: Absorbance of 1.0 at OD_{260} equivalent to 40 μg/ml of RNA.
5. Perform a MOPS gel electrophoresis for estimating the quality of the RNA preparations (see Subheading 3.1.1).
6. Synthesize the first strand cDNA by using High capacity cDNA reverse transcription kit (Applied Biosystems) as per the manufacturer's instructions. Use the cDNA thus synthesized for real-time PCR.
7. Set up the real-time PCR amplification by mixing 20 μl water, 2 μl 5′ primer (5 μM stock), 2 μl 3′ primer (5 μM stock), 1 μl cDNA, 25 μl SYBR Green PCR Master mix (Applied Biosystems).
8. Load 15 μl of this reaction mix in triplicate in Optical 96-well reaction plate (Applied Biosystems). *UBQ5* (ubiquitin) transcript can be used as an endogenous control for all the samples tested using the ubq5-specific forward and reverse primers.
9. Analyze the data thus obtained by the comparative C_T method (also referred to as the $2^{-\Delta\Delta C_T}$ method) (13).

4. Notes

1. Store the stock solutions of Yoshida's medium at 4°C. Stocks containing $CaCl_2$ and $FeCl_3$ should be checked after every 3 weeks as these tend to precipitate out.
2. Use RNase-free plasticware, glassware, and chemicals for RNA work by DEPC (diethylpyrocarbonate) treatment. Take care of the general precautions for working with RNA.
3. DEPC is a suspected carcinogen and should be handled with great care. Always wear gloves and use a fume hood when using this chemical.
4. TRIzol contains phenol and GITC which could be hazardous to humans. Always wear laboratory coat, gloves, and eye protection while handling this solution.
5. Stocks of MES and $MgCl_2$ can be stored at 4°C for 1 month.
6. Acetosyringone must always be prepared fresh (by dissolving in DMSO (dimethyl sulfoxide)) before every use.
7. Always store the stock solution and agar plates containing X-gal in dark as it is sensitive to light.
8. Store the stock solution, media, and agar plates containing rifampicin wrapped in aluminum foil as it is light sensitive. Always wear disposable gloves and safety glasses while handling the solution.
9. Do not incubate the LB agar plates containing antibiotic ampicillin at 37°C for long as it tends to give rise to satellite colonies. Also never use old ampicillin stocks or plates as the ampicillin tends to degrade, giving a reduced effective concentration.
10. Ethidium bromide (EtBr) is a potential carcinogen. While handling, always wear disposable gloves, eye protection, and lab coat. Avoid inhaling vapors. Agarose gels and buffers containing EtBr should be disposed of as per the recommended procedures.
11. During the agroinoculation procedure, take extreme care to avoid puncturing across the meristematic region/crown region of stem of the rice plants as this will lead to unsuccessful delivery of the vector.
12. Carefully discard the Whatman filter paper, syringes, disposable gloves, Yoshida's medium, microcentrifuge tubes, tissue papers, etc. used for the agroinoculation following proper procedures.
13. Cover the entire area used for agroinoculating rice seedlings properly with a blotting sheet to avoid spilling of the *Agrobacterium* suspension on the working area. Always wipe the area used for agroinoculation with ethanol afterward.

14. Cover the entire setup of the agroinoculated plants, kept for 24 h at 27°C, with an appropriately sized chamber to maintain the humidity.
15. Before transferring the agroinoculated plants into test tubes, after 24 h of agroinoculation, wash the seedlings thoroughly with water and pat dry to remove the excess *Agrobacterium* suspension sticking on to the roots of seedlings.
16. Remove all the dead and infected leaves from the agroinoculated plants regularly to avoid fungal infections. After every 20 days, wash the test tubes containing the agroinoculated rice plants and add fresh Yoshida's medium to avoid any algal growth.
17. Discard all the agroinoculated plants only after autoclaving at the recommended conditions.

Acknowledgments

AP and SS acknowledge the Research Fellowships from the University Grants Commission, New Delhi, and Council of Scientific and Industrial Research, New Delhi, respectively. This work was funded by the Department of Biotechnology, Government of India, New Delhi, Grant no. BT/AB/03/FG-I/2003 to ID.

References

1. Susi P, Hohkuri M, Wahlroos T et al (2004) Characteristics of RNA silencing in plants: similarities and differences across kingdoms. Plant Mol Biol 54:157–174
2. Szittya G, Molnar A, Silhavy D et al (2002) Short defective interfering RNAs of tombusviruses are not targeted but trigger post transcriptional gene silencing against their helper virus. Plant Cell 14:359–372
3. Moissiard G, Voinnet O (2006) RNA silencing of host transcripts by *Cauliflower mosaic virus* requires coordinated action of the four Arabidopsis Dicer-like proteins. Proc Natl Acad Sci USA 103:19593–19598
4. Purkayastha A, Dasgupta I (2009) Virus-induced gene silencing: a versatile tool for discovery of gene functions in plants. Plant Physiol Biochem 47:967–976
5. Holzberg S, Brosio P, Gross C et al (2002) *Barley stripe mosaic virus*-induced gene silencing in a monocot plant. Plant J 30:315–327
6. Ding XS, Schneider WL, Chaluvadi S et al (2006) Characterization of a *Brome mosaic virus* strain and its use as a vector for gene silencing in monocotyledonous hosts. Mol Plant Microbe Interact 19:1229–1239
7. Purkayastha A, Mathur S, Verma V et al (2010) Virus-induced gene silencing in rice using a vector derived from a DNA virus. Planta 232(6):1531–1540
8. International Rice Genome Sequencing Project (2005) The map-based sequence of the rice genome. Nature 436:793–800
9. Yoshida S, Forno SA, Cock SH et al (1976) Routine procedure for growing rice plants in culture solution. In: Yoshida S (ed) Laboratory manual for physiological studies of rice. IRRI, Manila, Philippines, pp 3367–3374
10. Sambrook J, Russell DW (2001) Molecular cloning: a laboratory manual, 3rd edn. Cold Spring Harbor Laboratory Press, Plainview, NY

11. Hood EE, Gelvin SB, Melchers LS et al (1993) New Agrobacterium helper plasmids for gene transfer to plants. Transgenic Res 2:208–218
12. Birnboim HC, Doly J (1979) A rapid alkaline extraction procedure for screening recombinant plasmid DNA. Nucleic Acids Res 7:1513–1523
13. Schmittgen DH, Livak JK (2008) Analyzing real-time PCR data by the comparative C_T method. Nat Protoc 3:1101–1108

Chapter 4

Utilizing Virus-Induced Gene Silencing for the Functional Characterization of Maize Genes During Infection with the Fungal Pathogen Ustilago maydis

Karina van der Linde and Gunther Doehlemann

Abstract

While in dicotyledonous plants virus-induced gene silencing (VIGS) is well established to study plant–pathogen interaction, in monocots only few examples of efficient VIGS have been reported so far. One of the available systems is based on the brome mosaic virus (BMV) which allows gene silencing in different cereals including barley (*Hordeum vulgare*), wheat (*Triticum aestivum*), and maize (*Zea mays*).

Infection of maize plants by the corn smut fungus *Ustilago maydis* leads to the formation of large tumors on stem, leaves, and inflorescences. During this biotrophic interaction, plant defense responses are actively suppressed by the pathogen, and previous transcriptome analyses of infected maize plants showed comprehensive and stage-specific changes in host gene expression during disease progression.

To identify maize genes that are functionally involved in the interaction with *U. maydis*, we adapted a VIGS system based on the *Brome mosaic virus* (BMV) to maize at conditions that allow successful *U. maydis* infection of BMV pre-infected maize plants. This setup enables quantification of VIGS and its impact on *U. maydis* infection using a quantitative real-time PCR (q(RT)-PCR)-based readout.

Key words: *Ustilago maydis, Zea mays* (maize), Virus-induced gene silencing, Brome mosaic virus (BMV), Plant–pathogen interaction

1. Introduction

Besides rice, maize (*Zea mays*) is not only one of the most important food resources of the increasing global human population but is also the major source for the production of biofuel. But significant global losses in crop production are caused by basidiomycete smut fungi which are one of the largest groups of plant pathogenic fungi (1). One of the best studied smut fungi is *Ustilago maydis*, the causative agent of corn smut disease which, due to its easy accessibility

Annette Becker (ed.), *Virus-Induced Gene Silencing: Methods and Protocols*, Methods in Molecular Biology, vol. 975, DOI 10.1007/978-1-62703-278-0_4, © Springer Science+Business Media New York 2013

to reverse genetic methods, has become a model system for biotrophic plant fungi (2). *U. maydis* is able to infect all aerial parts of the maize plant and transform them into tumor-like structures, within which the fungus completes its life cycle (3, 4). By using Affymetrix© maize microarrays, Doehlemann et al. (5) showed the regulation of an increasing number of maize genes throughout the infection process. To test which of these *U. maydis* responsive genes are functionally involved in the pathogen interaction, a functional test via gene silencing is required. However, in monocots, the production of stable RNA interference (RNAi) plants is a time-consuming and laborious procedure which requires large greenhouse capacities. A fast alternative to the generation of stable transgenic plants is virus-induced gene silencing (VIGS). All VIGS systems are based on the fact that plants defend themselves against RNA viruses while viral RNAs act as a trigger to induce RNA-mediated gene silencing, induced by sequence specific small RNAs (6). By inserting a fragment of a gene of interest (GOI) into the viral RNA, transcripts of this gene fragment become targets for degradation, resulting in a downregulation of the corresponding gene by posttranscriptional gene silencing (7, 8). VIGS has been successfully applied in other pathogen interactions with barley and wheat to demonstrate the functional requirement of particular genes for resistance (9–12). In 2006, Ding et al. (13) described silencing of the phytoene desaturase gene in rice, barley, and maize using a VIGS system based on the Brome mosaic virus (BMV). The genome of BMV encodes for three RNAs, which were cloned as cDNAs resulting in the vectors pF1-11, pF2-2, and pB3-3 (13, 14). For silencing, a fragment of the GOI can be inserted into the 3′-UTR of cDNA 3 (pB3-3). Inoculation of maize plants with transcripts from pF1-11, pF2-2, and pB3-3 bearing a fragment of the GOI should cause silencing of this gene.

We adapted the BMV system for the use in the *U. maydis*–maize pathosystem. Therefore, we standardized all steps of the described system to ensure comparable virus infections and subsequent silencing rates. In addition, a quantitative real-time PCR-based readout was applied to facilitate the direct correlation of silencing efficiency and colonization of the plant by *U. maydis.*

Since in vitro transcription of viral RNAs is relatively expensive, viral transcripts are first inoculated on *Nicotiana benthamiana* leaves (Fig. 1). Tobacco is easily infected and accumulates high titers of virus (14). To guarantee comparable virus titers in all tobacco extracts for VIGS experiments, virus titers in tobacco sap can be determined by qPCR and subsequently adjusted before inoculation on maize (Fig. 1). Since insertion of a gene fragment in RNA3 results in decreased BMV pathogenicity (15), all silencing approaches should also be performed with a VIGS construct containing a fragment of a non-plant gene (e.g., for the yellow fluorescent protein *yfp*). This practice has been also proposed

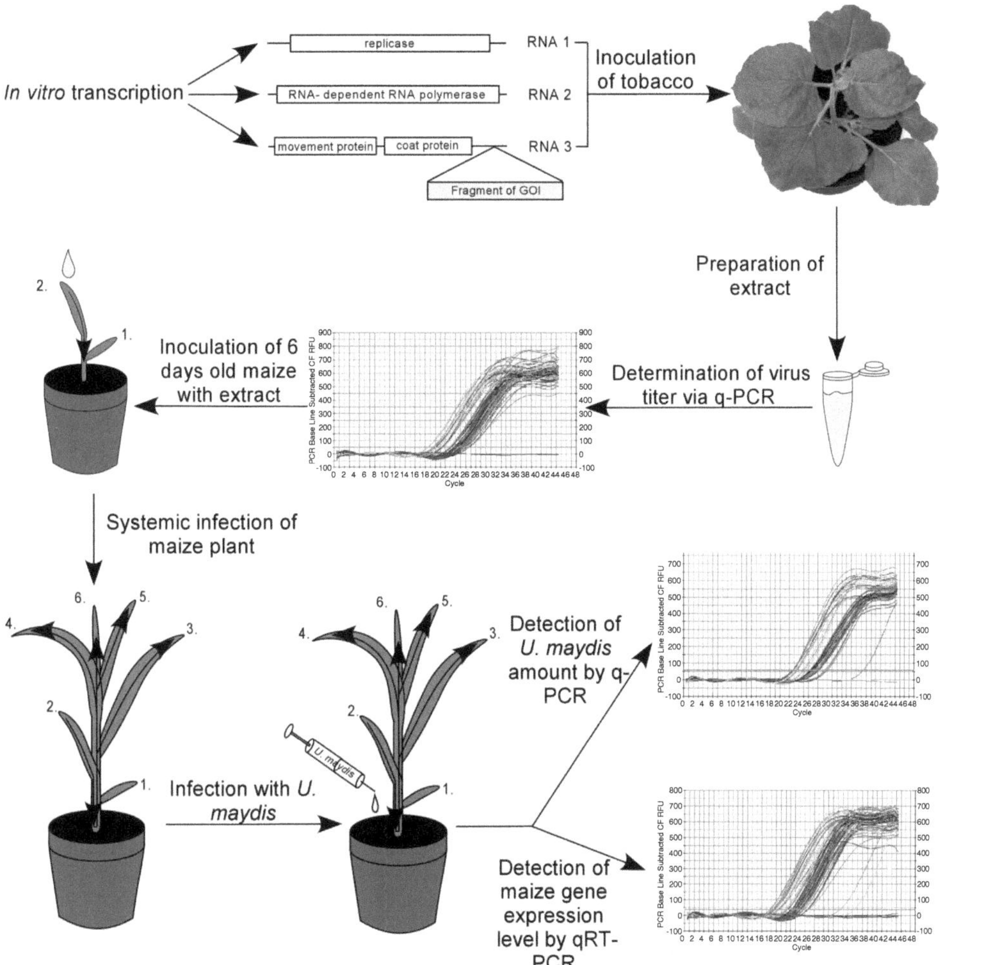

Fig. 1. Experimental setup for a quantitative readout to investigate the influence of VIGS in maize during the interaction with *Ustilago maydis*. GOI, gene of interest.

previously by Scofield and Nelson (7). After inoculation of the second maize leaf with the tobacco extract containing the viral RNAs, the virus spreads systemically within the maize plant (Fig. 1), resulting in silencing of the GOI in leaves four, five, and six, while strong, reproducible silencing only occurs in leaf six (15). However, also in leaf six not all cells silence the GOI. Silencing generally occurs in clusters of cells, often in the vicinity of vascular bundles (15). After infection of leaf six with *U. maydis*, silencing efficiency and fungal biomass can be correlated by probing BMV- and *U. maydis*-infected leaf material of individual plants by q(RT)-PCR (Fig. 1). Testing individual plants is essential, since silencing rates differ from plant to plant even if all conditions are adjusted and standardized.

By establishing a quantitative readout for VIGS in maize that allows parallel infection with *U. maydis*, we were able to develop a versatile tool to study the effects of plant genes on the pathogen interaction. Here, we provide a detailed description of all steps required for silencing of maize genes using the BMV system.

2. Materials

2.1. Cloning of Maize Sequences in pB3-3

1. BioMix Red (Bioline, Luckenwalde, Germany).
2. 1×TAE: Dissolve 4.84 g Tris-HCL in 500 ml H_2O, add 2 ml 0.5 M Na_2EDtA (pH 8.0) and 1.14 ml glacial acetic acid; adjust to 1 l with H_2O.
3. 2% TAE-agarose gel: Weigh out 2 g of agarose (Biozym, Vienna, Austria) into a flask and add 100 ml of 1× TAE. Heat until agarose is completely dissolved. Cool down to 60°C and add 10 μl ethidium bromide (Roth, Karlsruhe, Germany) and pour into gel tray.
4. 6× agarose-loading buffer: Dissolve 150 mg bromophenol blue in 33 ml 150 mM Tris-HCl pH 7.6. Add 60 ml glycerol and 7 ml H_2O.
5. Wizard SV Gel and PCR Clean-Up System (Promega, Mannheim, Germany).
6. Plasmid pB3-3.
7. *Hin*dIII, buffer 2 (New England Biolabs, Frankfurt am Main, Germany).
8. Antarctic phosphatase buffer, Antarctic phosphatase (New England Biolabs, Frankfurt am Main, Germany).
9. T4 Ligase, 10× T4 Ligase buffer (New England Biolabs, Frankfurt am Main, Germany).
10. Chemical competent *Escherichia coli* DH5α cells (Invitrogen, Darmstadt, Germany).
11. dYT: Dissolve 16 g Tryptone–Peptone (Difco, Lawrence, Kansas, USA), 10 g yeast extract (Difco, Lawrence, Kansas, USA), and 5 g NaCl (Roth, Karlsruhe, Germany) in 1,000 ml H_2O. Sterilize by autoclaving.
12. YT agar: Dissolve 6.4 g Tryptone–Peptone (Difco, Lawrence, Kansas, USA), 4 g yeast extract (Difco, Lawrence, Kansas, USA), and 4 g NaCl (Roth, Karlsruhe, Germany) in 800 ml H_2O. Add 10.4 g agar (Roth, Karlsruhe, Germany). Sterilize by autoclaving. Cool to 50°C, add 8 ml of 10 mg/ml ampicillin, and pour into sterile plates.
13. Buffer P1, buffer P2, buffer P3 (Qiagen, Hilden, Germany).
14. Isopropanol, 70% ethanol.

2.2. Synthesis of BMV RNAs

1. Plasmids pF1-11, pF2-2.
2. *Psh*AI, *Spe*I, buffer 4, 100× BSA (New England Biolabs, Frankfurt am Main, Germany).
3. Wizard SV Gel and PCR Clean-Up System (Promega, Mannheim, Germany).
4. mMESSAGE mMACHINE T3 Kit (Ambion, Austin, Texas, USA).
5. 1% TAE-agarose gel: Weigh out 1 g of agarose (Biozym, Vienna, Austria) into a flask and add 100 ml of 1× TAE. Heat until agarose is completely dissolved. Cool down to 60°C, add 10 μl ethidium bromide (Roth, Karlsruhe, Germany), and pour into gel tray.
6. 6× agarose-loading buffer: (for preparation, see Subheading 2.1).

2.3. Inoculation of Tobacco with In Vitro Transcribed RNAs

1. *N. benthamiana* seeds.
2. Pikiererde type P soil (Fruhstorfer, Vechta, Germany).
3. Fine brush.
4. Carborundum 400 mesh (Sigma, Munich, Germany).

2.4. Preparation of BMV Tobacco Extract and Virus Titer Detection

1. 100 mM sodium phosphate buffer: For 0.2 M sodium phosphate, monosodium salt, dissolve 27.6 g NaH_2PO_4-$1H_2O$ (Roth, Karlsruhe, Germany) in 1 l H_2O. For 0.2 M sodium phosphate, disodium salt, dissolve 53.62 g Na_2HPO_4-$7H_2O$ in 1 l H_2O. Mix 87.7 ml sodium phosphate monosalt solution, 12.3 ml sodium phosphate disalt solution, and 100 ml H_2O.
2. Mortar and pistil.
3. iQ SYBR Green Supermix (BioRad, Munich, Germany), Fluorescein (BioRad, Munich, Germany).

2.5. Inoculation of Maize cv. Va35 with BMV

1. *Z. mays* cv. Va35 seeds.
2. "Pikiererde" type P soil (Fruhstorfer, Vechta, Germany).
3. Fine brush.
4. Carborundum 400 mesh (Sigma, Munich, Germany).

2.6. U. maydis Infection of Maize

1. Yepsl: Dissolve 10 g yeast extract (Difco, Lawrence, Kansas, USA), 4 g Bacto-Peptone (Difco, Lawrence, Kansas, USA), and 4 g sucrose (Roth, Karlsruhe, Germany) in 1 l H_2O and sterilize by autoclaving.
2. *U. maydis* solopathogenic strain SG200.
3. 1-ml syringe and a 22-gauge needle.

2.7. Readout by q(RT)-PCR

1. 1% Tween-20: Add 10 ml Tween-20 (Roth, Karlsruhe, Germany) to 990 ml H_2O.

2. Reaction-tube pistil.
3. MasterPure Complete DNA and RNA Purification Kit (Epicenter Biotechnologies, Wisconsin, USA).
4. 1% TAE-agarose gel: (for preparation, see Subheading 2.2).
5. First Strand cDNA synthesis Kit (Fermentas, St. Leon, Germany).
6. iQ SYBR Green Supermix (BioRad, Munich, Germany), Fluorescein (BioRad, Munich, Germany).

3. Methods

3.1. Cloning of Foreign Maize Sequences in pB3-3

1. Choose fragments of GOI with a length of 150–250 bp.
2. Enter the sequences in siRNA Scan (http://bioinfo2.noble.org/RNAiScan.htm) (see Note 1).
3. Use the fragment which gives most siRNAs.
4. Construct primers containing a *Hin*dIII restriction site and CCC extension at the 5′ end to amplify this fragment from maize cv. Va35 cDNA.
5. Amplify the fragment in a total volume of 50 μl with 25 μl of 2× BioMix Red (Bioline, Luckenwalde, Germany), 1 μl 10 pmol/μl forward primer, 1 μl 10 pmol/μl reverse primer, and 1 μl 50 ng/μl maize cv. Va35 cDNA.
6. Apply PCR on a 2%-TAE-agarose gel and extract the PCR product from the gel.
7. Purify the PCR product by using Wizard SV Gel and PCR Clean-Up System (Promega, Mannheim, Germany) according to the manufactures' protocol.
8. Digest complete amount of PCR product and 1 μg of pB3-3 overnight at 37°C each in a total volume of 50 μl with 1 μl *Hin*dIII and 5 μl 10× buffer 2.
9. Purify both restriction reactions with Wizard SV Gel and PCR Clean-Up System (Promega, Mannheim, Germany) according to the manufactures' protocol and elute in 26 μl nuclease-free water.
10. Add 3 μl Antarctic phosphatase buffer and 1 μl Antarctic phosphatase to the purified restriction reaction of pB3-3 and incubate for 30 min at 37°C; afterward, heat inactivate the reaction for 5 min at 65°C.
11. Ligate pB3-3 and PCR product (molecular ration 1:1) with 1 μl T4 Ligase and 2 μl 10× T4 Ligase buffer overnight at 16°C in a total volume of 20 μl.

12. Add 20 μl of ligation reaction to 50 μl competent *E. coli* DH5α cells (Invitrogen, Darmstadt, Germany), mix by pipetting up and down three times, incubate 30 min on ice, heat shock 1 min at 42°C, add 200 μl of dYT medium, incubate 1 h at 37°C by shaking at 200 rpm, streak out on YT-agar plates supplemented with ampicillin, and incubate for 16 h at 37°C.
13. Pick single colonies, streak out on YT-agar plates supplemented with ampicillin, and incubate again for 16 h at 37°C.
14. To screen single bacterial colonies by colony PCR, colonies are picked using a sterile pipette tip, dipping it in a PCR tube containing 10 μl 2× BioMix Red (Bioline, Luckenwalde, Germany), 1 μl 10 pmol/μl primer 5′-GAGCCCCTGACTGGGTTAAAG-3′, 1 μl 10 pmol/μl forward primer (specific to your fragment of interest,) and 8 μl H_2O. Amplify 35 cycles (see Note 2).
15. Apply PCRs on 2% TAE-agarose gel. Colonies from which a PCR product is obtained are inoculated with 5 ml dYT medium for 16 h at 37°C, constantly shacking at 200 rpm.
16. For plasmid preparation and subsequent sequencing, harvest 4 ml bacterial cells by centrifugation at 13,000 × *g* for 1 min. Resuspend the bacterial pellet in 0.3 ml of buffer P1 (Qiagen, Hilden, Germany). Add 0.3 ml of buffer P2 (Qiagen, Hilden, Germany); mix thoroughly by vigorously inverting the tube 4–6 times. Add 0.3 ml of buffer P3 (Qiagen, Hilden, Germany); mix immediately and thoroughly by inverting 4–6 times. Centrifuge at maximum speed in a microcentrifuge for 10 min. Pipette 600 μl of clear supernatant containing plasmid DNA in a new tube and add 600 μl 100% isopropanol. Mix by inverting the tube 4–6 times and centrifuge at maximum speed in a microcentrifuge for 10 min. Discard supernatant and add 800 μl of 80% ethanol to the pellet. Centrifuge at maximum speed in a microcentrifuge for 2 min, discard supernatant, and dry pellet for 5 min at RT. Resuspend the pellet in 50 μl of nuclease-free water.
17. Sequence plasmid with primer 5′-GAGCCCCTGACTGGGTT AAAG-3′ (same primer as step 14).

3.2. Synthesis of BMV RNAs

All reagents and materials used should be nuclease-free to avoid degradation of RNAs (see Note 3).

1. Incubate 3 μg pF1-11, pF2-2, and pB3-3 with fragment of interest or fragment of *yfp* in a 50 μl digest reaction with 1.5 μl *Psh*AI or *Spe*I and 5 μl 10× Restriction buffer for 1.5 h at 37°C (see Note 4).
2. After incubation, add 50 μl of Wizard SV Gel and PCR Clean-Up System membrane binding solution to total restriction reaction and apply to a Wizard purification column. Purify

restriction according to Wizard SV Gel and PCR Clean-Up System (Promega, Mannheim, Germany) manual. Elute linearized plasmids with 30 μl of nuclease-free water.

3. Evaporate down to a volume of 2–6 μl in vacuum under rotation at RT.
4. For in vitro transcription, thaw all frozen reagents of the mMESSAGE mMACHINE T3 Kit (Ambion, Austin, TX, USA), mix vigorously, and spin down. Add 10 μl 2× NTP/CAP, 2 μl 10× reaction buffer, complete amount of restriction reaction, and fill with nuclease-free water up to 18 μl (all from mMESSAGE mMACHINE T3 Kit, Ambion, Austin, TX, USA). Start reaction by adding 2 μl T3 enzyme mix (mMESSAGE mMACHINE T3 Kit, Ambion, Austin, TX, USA); mix and incubate 2 h at 37°C.
5. Add 1 μl TURBO DNase (Ambion, Austin, TX, USA) and mix well. Incubate at 37°C for 15 min.
6. For estimation of in vitro transcription quality and quantity, mix 1 μl of in vitro transcription with 2 μl 6× agarose-loading buffer and 7 μl nuclease-free H_2O and apply on a 1% TAE-agarose gel (see Note 5).
7. RNAs can be stored for short terms at –20°C and for longer terms at –80°C.

3.3. Inoculation of Tobacco with In Vitro Transcribed RNAs

While in our hands it was not possible to successfully inoculate maize with the three RNAs directly, we therefore amplified the virus first in *N. benthamiana*.

1. Sow about 40 *N. benthamiana* seeds in a 9×9 cm pot containing Pikiererde type P soil (Fruhstorfer, Vechta, Germany) and grow 3 weeks at 28/22°C (day/night) with 14.5 h light period and 1 h of sunset and sunrise.
2. Afterward, individual seedlings are pricked in 9×9 cm pots and grown again for 2 weeks under the above described conditions.
3. One day before BMV inoculation, plants are moved into a climate chamber with 22°C during the light period (26,000 lux, 14.5 h) and 20°C during the dark period (9.5 h) (see Note 6).
4. Mix the three RNAs (RNA1, RNA2, and RNA3) with fragment of interest or RNA3 containing a fragment of *yfp* (see Note 6) in an equimolar ratio and a total volume of 7.5 μl.
5. At the beginning of the dark period, two leaves of one tobacco plant are dusted with a fine brush with Carborundum 400 mesh (Sigma, Munich, Germany), then 3.75 μl of premixed RNAs are spotted on the leaf surface and rubbed in with a glove-coated finger (see Note 7).

3.4. Preparation of BMV Tobacco Extract and Virus Titer Detection

1. 5–6 days after inoculation of the tobacco plants, harvest the inoculated leaves as well as the upper systemic infected leafs showing mosaic symptoms.
2. Weigh the leaves and grind with 100 mM sodium phosphate buffer, pH 6.0 (100 μl buffer/10 mg leaf material), in a mortar.
3. Aliquot the extracts and store at −20°C.
4. For virus titer detection, thaw one aliquot, centrifuge at maximum speed in a microcentrifuge for 2 min, and transfer the supernatant into a new tube.
5. Prepare 1:10 and 1:100 dilutions of the tobacco supernatants in 100 mM sodium phosphate buffer, pH 6.0.
6. For qPCR, mix in a mastermix for each reaction 12.5 μl iQ SYBR Green Supermix (BioRad, Munich, Germany), 0.5 μl 10 pmol/μl primer BMV-RNA1-RT-for 5′-ATATGGGCTTCCAAGGTC TG-3′, 0.5 μl 10 pmol/μl primer BMV-RNA1-RT-rev 5′-TC GCGGTCAAACAACATGGC-3′, 0.5 μl 1 μM Fluorescein (BioRad, Munich, Germany), and 10 μl H_2O. In a second mastermix, add for each reaction 0.5 μl 10 pmol/μl primer Actin-RT-for 5′-GGAATCCACGAGACCACATA-3′ and 0.5 μl 10 pmol/μl primer Actin-RT-rev 5′-CATTCTA TCGGCAATACCTGG-3′ (see Note 8). For each template, prepare two qPCR reaction tubes with 24 μl BMV-RNA1-RT primers mastermix and two tubes with Actin-RT primers mastermix. Add 1 μl of template (tobacco extract, tobacco extract dilutions, or H_2O) to all tubes, vortex briefly, and spin down.
7. Perform qPCR with cycling conditions as follows: 2 min at 95°C, followed by 45 cycles of 30 s at 95°C, 30 s at 61°C, and 30 s at 72°C and 70°C dissociation curve starting temperature. Afterward, check all dissociation curves to make sure that every reaction just has a single peak. Check amplification curve to make sure there is no product formed for the H_2O control. Set up proper threshold to get the Ct values in the range of exponential amplification. Export Ct values to Rest2009 software (Qiagen, Hilden, Germany) and analyze, while amount of BMV RNA1 in uninfected tobacco is set to one.

3.5. Inoculation of Maize cv. Va35 with BMV

1. Sow 4 *Z. mays* cv. Va35 seeds in each round 15 cm pot containing Pikiererde type P soil (Fruhstorfer, Vechta, Germany) and grow 5 days at 28/22°C (day/night) with 14.5 h light period and 1 h of sunset and sunrise.
2. On day 5 after potting plants are moved into a climate chamber with 28°C during the light period (26,000 lux, 14.5 h) and 22°C during the dark period (9.5 h).

3. Adjust virus titer in all samples to the same amount of BMV RNA1 according to your qPCR results (see Note 9).
4. 6 days after potting, at the beginning of the dark period, the second leaf of maize is dusted with a fine brush with Carborundum 400 mesh (Sigma, Munich, Germany), then 50 μl of tobacco extract is spotted on the leaf surface and rubbed in with a glove-coated finger.

3.6. *U. maydis* Infection of Maize

1. 16 days after potting the maize plants, inoculate 5 ml of Yepsl with *U. maydis* solopathogenic strain SG200 and incubate for 24 h at 28°C and constantly shacking at 200 rpm.
2. Dilute well-grown SG200 overnight culture to an OD_{600nm} of 0.2 and grow again for 4 h. Harvest cells by centrifugation at 900 × g for 5 min, discard supernatant, and resuspend cell pellet in H_2O to an OD_{600nm} of 1.
3. At the beginning of the light period, 17 days after potting (11 days after BMV inoculation), the freshly prepared *U. maydis* suspension is injected in the leaf shield about 2 cm above ground with a 1-ml syringe and a 22-gauge needle.

3.7. Readout by q(RT)-PCR

1. Depending on the expression maximum of your gene harvest, plant material by cutting the whole plant 0.5 cm above ground and separate leaf 6.
2. Wipe leaf number 6 three times with a paper towel wetted with 1% Tween-20 and three times with a water-wetted paper towel (see Note 10).
3. Cut the leaf 1 and 3 cm below the injection holes and directly freeze the released 2 cm of leaf in a reaction tube in liquid nitrogen (see Note 11). Harvest 10 leaves of BMV/YFPsi-inoculated plants infected with *U. maydis* and 10 leaves of plants inoculated with BMV to silence your GOI and infected with *U. maydis* in individual reaction tubes.
4. Grind the leaf in liquid nitrogen directly in the reaction tube with a reaction-tube pistil.
5. Prepare DNA and RNA by using the MasterPure Complete DNA and RNA Purification Kit (Epicenter Biotechnologies, Wisconsin, USA). Add 300 μl Tissue and Cell Lysis solution and 1 μl Proteinase K, mix well, incubate 15 min at 65°C, centrifuge 2 min at maximum speed in a microcentrifuge, transfer the supernatant into a fresh reaction tube, incubate again for 15 min at 65°C, and cool down on ice. Aliquot half of the reaction in a fresh reaction tube. To one half, add 0.5 μl RNase A, mix, incubate 30 min at 37°C, and cool down on ice (this half is for DNA preparation). Then add 75 μl MPC Protein Precipitation Reagent, mix, and centrifuge at maximum speed in a microcentrifuge for 10 min. Transfer supernatant into a

new reaction tube, and add 500 µl 100% isopropanol; invert tubes and centrifuge 15 min 17,000×*g* at 4°C. Take off the isopropanol carefully and wash 1 time with 75% ethanol. Dry pellet 5 min and resuspend in 35 µl TE buffer. To the other half, add 75 µl MPC Protein Precipitation Reagent, mix, and centrifuge at maximum speed in a microcentrifuge for 10 min (this one is for RNA preparation). Transfer the supernatant into a new reaction tube and add 500 µl of 100% isopropanol. Invert the tube and centrifuge 15 min 17,000×*g* at 4°C. Take off the isopropanol carefully and resuspend the pellet in 100 µl DNase I buffer and 2.5 µl DNase I. Incubate at 37°C for 30 min, then add 100 µl 2× T and C Lysis solution and vortex. Add 100 µl of MPC Protein Precipitation Reagent, vortex, and place on ice for 5 min. Centrifuge at maximum speed in a microcentrifuge for 10 min. Transfer the supernatant to a new tube and precipitate by adding 500 µl isopropanol. Wash the pellet once with 75% ethanol and dry. Resuspend RNA in 35 µl TE and 1 µl ScriptGuard.

6. Check RNA quality by applying 1 µl mixed with 2 µl 6× agarose-loading buffer and 7 µl nuclease-free H_2O on a 1% TAE-agarose gel, and determine the RNA amount by measuring absorption at 260 nm (see Note 12).
7. For cDNA synthesis with the First Strand cDNA synthesis Kit (Fermentas, St. Leon, Germany), pipette 0.5 µl oligo dT primer, 2 µl 5× reaction buffer, 0.5 µl RiboLock RNase inhibitor, 1 µl 10 mM dNTP mix, 1 µl M-MuLV reverse transcriptase, and 0.5 µg of RNA into a reaction tube, and fill up to a total volume of 10 µl with nuclease-free water. Incubate 60 min at 37°C and 5 min at 70°C (see Note 13).
8. To check for silencing of your GOI, perform qRT-PCR on cDNA with gene-specific primers and primers for *gapdh* (Ncbi gene bank NM001111943) as reference. Mix in a mastermix for each reaction 12.5 µl iQ SYBR Green Supermix (BioRad, Munich, Germany), 0.5 µl 10 pmol/µl gene-specific forward primer, 0.5 µl 10 pmol/µl gene-specific reverse primer (see Note 14), 0.5 µl 1 µM Fluorescein (BioRad, Munich, Germany), and 10 µl H_2O. In a second mastermix, add for each reaction 0.5 µl 10 pmol/µl primer GAPDH-RT-for 5-CTTCGGCATTGTTGAGGGTTTG-3 and 0.5 µl 10 pmol/µl primer GAPDH-RT-rev 5-TCCTTGGCT GAGGGTCCGTC-3. For each template, prepare two qPCR reaction tubes with 24 µl gene-specific primers mastermix and two tubes with GAPDH-RT primers mastermix. Add to all tubes 1 µl of template (cDNA or H_2O), vortex briefly, and spin down. Perform qRT-PCR as described in 3.4.7.
9. For the determination of *U. maydis* amount *in planta,* perform qPCR on 1 µl of DNA with primer for *U. maydis* peptidyl-

prolyl *cis–trans* isomerase (*ppi*, Um03726) Um03726-RT-for 5′-ACATCGTCAAGGCTATCG-3′ and Um03726-RT-rev 5′-AAAGAACACCGGACTTGG-3′ and GAPDH-RT primers (see step 7) as described in Subheading 3.4.

10. After qRT-PCR on RNA and qPCR on DNA, check all dissociation curves to make sure that every reaction just has a single peak. Check amplification curve to make sure there is no product formed for the H_2O control. Set up proper threshold to get the Ct values in the range of exponential amplification. Export Ct values to Rest2009 software (Qiagen, Hilden, Germany) and analyze. Expression of your GOI in BMV/YFPsi-inoculated plants infected with *U. maydis* as well as the amount of *ppi* DNA (amount of *U. maydis*) in BMV/YFPsi-inoculated plants infected with *U. maydis* are set to one. Now you can compare *U. maydis* amounts in plants which silence your GOI and non-silencing plants.

4. Notes

1. Use the preset specifications and choose the maize database.
2. By using the cloning forward primer in combination with the primer binding 82 bp upstream of the *Hin*dIII side, only reverse inserted fragments are amplified. In our hands, reverse inserted fragments lead to higher silencing rates than forward inserted fragments.
3. To avoid RNase contaminations, use gloves and filter-pipette tips when working with RNA. Furthermore, work fast, if possible on ice, and spin down everything before opening a tube to preserve RNA from degradation.
4. pF1-11 is linearized with *Spe*I, while *Psh*AI is used for pF2-2 and all pB3-3 constructs.
5. Quantity of RNA can be determined by comparison with marker lanes when defined amounts of marker were loaded to the gel.
6. Remember to put also one control tobacco plant in the climate chamber which will not be inoculated with BMV RNAs.
7. Do not squish the tobacco leaf; it suffices to rubber carefully till a slight smell of tobacco comes out of the leaf.
8. Before qPCR, boil primers for 5 min and then put them directly on ice; this avoids the formation of secondary structures by the primers.
9. For us, a virus titer of 10,000 relative expression units works fine, but in our opinion, in every lab this titer varies. To find the optimal titer, inoculate maize plants with different dilutions

of tobacco extract challenged with BMV/PDSsi and check for best photo bleaching results.

10. To remove non-penetrated *U. maydis* cells from the leaf surface, completely make sure to wipe the whole leaf from both sides.
11. Try to work as fast as possible from the point you cut the plant till the leaf part is frozen in liquid nitrogen to avoid unexpected side effects resulting from cutting the leaf.
12. An absorption of 1 at 260 nm corresponds to an RNA amount of 40 ng/μl.
13. After cDNA synthesis, place cDNA direct on ice and use for qRT-PCR or store at -20°C (for less than 1 week). For long-term storage, -80°C is recommended.
14. Choose a primer pair binding relatively close to the Stop codon of you GOI, having a melting temperature of 60°C that will amplify a fragment with a length of 80–120 bp. Both primers should have a length between 18 and 22 bp.

Acknowledgments

We thank Alexander Hof for helpful comments on the manuscript. The work was funded by the Max Planck Society and the DFG research group FOR666.

References

1. Martinez-Espinoza AD, Garcia-Pedrajas MD, Gold SE (2002) The Ustilaginales as plant pests and model systems. Fungal Genet Biol 35:1–20
2. Kämper J, Kahmann R, Bölker M et al (2006) Insights from the genome of the biotrophic fungal plant pathogen *Ustilago maydis*. Nature 444:97–101
3. Brefort T, Doehlemann G, Mendoza-Mendoza A et al (2009) *Ustilago maydis* as a pathogen. Annu Rev Phytopathol 47:423–445
4. Skibbe DS, Doehlemann G, Fernandes J et al (2010) Maize tumors caused by *Ustilago maydis* require organ-specific genes in host and pathogen. Science 328:89–92
5. Doehlemann G, Wahl R, Horst R et al (2008) Reprogramming a maize plant: transcriptional and metabolic changes induced by the fungal biotroph *Ustilago maydis*. Plant J 56:181–195
6. Purkayastha A, Dasgupta I (2009) Virus-induced gene silencing: a versatile tool for discovery of gene functions in plants. Plant Physiol Biochem 47:967–976
7. Scofield SR, Nelson R (2009) Resources for virus-induced gene silencing (VIGS) in the grasses. Plant Physiol 149:152–157
8. Ratcliff F, Harrison BD, Baulcombe DC (1997) A similarity between viral defense and gene silencing in plants. Science 276:1558–1560
9. Scofield SR, Huang L, Brandt AS et al (2005) Development of a virus-induced gene silencing system for hexaploid wheat and its use in functional analysis of the Lr21-mediated leaf rust resistance pathway. Plant Physiol 138:2165–2173
10. Zhou H, Li S, Deng Z et al (2007) Molecular analysis of three new receptor-like kinase genes from hexaploid wheat and evidence for their participation in the wheat hypersensitive response to stripe rust fungus infection. Plant J 52:420–434
11. Hein I, Barciszewska-Pacak M, Hrubikova K et al (2005) Virus-induced gene silencing-based functional characterization of genes associated with powdery mildew resistance in barley. Plant Physiol 138:2155–2164

12. Sindhu A, Chintamanani S, Brandt AS et al (2008) A guardian of grasses: specific origin and conservation of a unique disease-resistance gene in the grass lineage. Proc Natl Acad Sci USA 105:1762–1767
13. Ding XS, Schneider WL, Chaluvadi SR et al (2006) Characterization of a Brome mosaic virus strain and its use as a vector for gene silencing in monocotyledonous hosts. Mol Plant Microbe Interact 19:1229–1239
14. Ding XS, Chaluvadi SR, Nelson RS (2007) Analysis of gene function in rice through virus-induced gene silencing. Methods Mol Biol 354:145–160
15. van der Linde K, Kastner C, Kumlehn J et al (2011) Systemic virus-induced gene silencing allows functional characterization of maize genes during biotrophic interaction with *Ustilago maydis*. New Phytol 189:471–483

Chapter 5

Analysis of Developmental Control Genes Using Virus-Induced Gene Silencing

Koen Geuten, Tom Viaene, Dries Vekemans, Sofia Kourmpetli, and Sinead Drea

Abstract

A consistent challenge in studying the evolution of developmental processes has been the problem of explicitly assessing the function of developmental control genes in diverse species. In recent years, virus-induced gene silencing (VIGS) has proved to be remarkably adaptable and efficient in silencing developmental control genes in species across the angiosperms. Here we describe proven protocols for *Nicotiana benthamiana* and *Papaver somniferum*, representing a core and basal eudicot species.

Key words: Virus-induced gene silencing, Plant development, TRV vector, RNA silencing

1. Introduction

In recent years, it has become clear that virus-induced gene silencing (VIGS) can be used as a technique to analyze the function of genes involved in plant development. This was unexpected, initially, due to the inability of many viruses to penetrate meristematic tissues, and meristem culture had been used routinely to obtain virus-free germplasm for many species. The first species in which VIGS was successfully used to obtain gene-silenced phenotypes in flower development was *Nicotiana benthamiana* (1, 2), and the method was subsequently adapted to silence genes involved in developmental processes in other species across the angiosperms. Using VIGS to analyze developmental control genes has both advantages and drawbacks in comparison to using stable transformation to investigate gene function. Advantages are that the method is (1) applicable to species for which no stable transformation protocol is available, (2) it is possible

Annette Becker (ed.), *Virus-Induced Gene Silencing: Methods and Protocols*, Methods in Molecular Biology, vol. 975,
DOI 10.1007/978-1-62703-278-0_5, © Springer Science+Business Media New York 2013

to analyze genes with an otherwise lethal phenotype, and (3) obtaining phenotypes can be exceptionally rapid and easy. Conversely, (1) efficiency can be low and variable depending on the species, (2) only partial phenotypes are obtained in many species, (3) sometimes more extensive phenotypic analysis is needed, and (4) silencing is not transmitted to the next generation (but see reference 3). This makes it virtually impossible to investigate genes involved in early processes of plant development (e.g., embryo development, juvenile–adult transition). What determines which developmental processes can be investigated is their timing and on which point during that process silencing can be induced. To induce optimal silencing early in development, different methods can be used, and we will describe two methods in two species. In addition, both the timing of developmental transitions and the efficiency of VIGS are strongly dependent on environmental factors, such as light regime and intensity, temperature, and humidity, and these factors may have to be optimized for the specific growth conditions used.

We routinely use *N. benthamiana* to analyze genes involved in development (4). *N. benthamiana* provides a system complementary to other Solanaceae species, such as *Petunia*, tobacco (*Nicotiana tabacum*), and tomato, and has as a main advantage that VIGS symptoms in this species are generally very pronounced and persistent (5). This probably results from the natural loss-of-function of an RNA-dependent RNA polymerase in *N. benthamiana* (6, 7). Forward genetic screens have already been performed using VIGS in this species (5, 8). Furthermore, this species can be stably transformed to complement gene-silencing phenotypes with overexpression phenotypes (9). Although *N. benthamiana* is a preferred model system to investigate plant virology (10), little has been published on its development despite the importance of this information when applying VIGS to silence developmental control genes. To overcome this problem, we determined the relative timing of developmental transitions using both phenotypic and molecular markers (Viaene and Geuten, unpublished). Our observations indicate that in long-day grown plants at 25°C (in a tightly controlled environment like a growth chamber), the juvenile–adult transition starts before the third leaf develops visibly, flowering is induced before the eighth leaf develops, and flower development initiates before the 11th leaf develops visibly. This means that by using leaf infiltration as an inoculation method, genes involved in the juvenile–adult transition cannot be studied, while floral induction is one of the earliest developmental processes that can be investigated. Yet using vacuum infiltration of *N. benthamiana* seedlings using the protocol described here for *Papaver* can make it possible to study earlier developmental processes. Growing these plants in less controlled environments like greenhouses alters the timing and details of these developmental transitions and should be reexamined for specific conditions. Under the previously defined conditions, a new leaf emerges approximately every 2 days, and as in other species of

Nicotiana, the number of leaves can be used as a proxy for flowering time. The first flower develops on the main stem and, even in wild-type plants, often shows developmental defects. Therefore, Observations for this flower cannot be interpreted as phenotypes that result from gene silencing. Inflorescence development in *N. benthamiana* is sympodial, which means that the inflorescence meristem commits to flower development when a new inflorescence meristem develops in a lower position on the axis. The *N. benthamiana* inflorescence rarely shows secondary branching. Descriptions of wild-type flower development can be found (2, 4). Protocols for VIGS in *N. benthamiana* have been previously described in detail (11, 12) Here we describe how we apply these protocols.

Outside of the core eudicots, VIGS has been adapted efficiently for several species in the basal eudicot group of angiosperms and in the Ranunculales order specifically. Initial success in *Papaver somniferum*, with detailed description of the method below, was followed by the development of efficient methods in *Eschscholzia california*, *Aquilegia*, and *Thalictrum* (13–17). This permits an opportunity for more localized gene function comparisons within the basal eudicots, as well as broader comparisons to orthologous gene function in core eudicots and monocots. VIGS for these basal species was adapted for classic evo–devo studies, and most of the genes examined to date have been orthologues of classic developmental regulator genes shown to have significant developmental defects in model plants such as *Arabidopsis*, *Antirrhinum*, *Petunia*, and maize or rice in the monocots. In addition, the genes analyzed using VIGS in these species have been mainly involved in various aspects of flower development. It remains to be seen if VIGS would be as effective in ascertaining the function of genes with more subtle effects on development and careful examination of plant material would be even more important in these cases. Like *N. benthamiana*, *P. somniferum* was not initially used as a model for developmental studies, but mostly regarded as a host to unique alkaloid production pathways resulting in the synthesis of important pharmaceuticals such as morphine and codeine (18, 19). However, data and resources developed for these more applied purposes are often a good starting point for adopting a species for more fundamental developmental research.

2. Materials

2.1. Vectors and Agrobacterium Strains

1. Vectors pTRV1 and pTRV2 necessary for tobacco rattle virus-based VIGS can be obtained from the ABRC stock center (Ohio State University).

2. Electrocompetent *A. tumefaciens* strain GV3101 for either *P. somniferum* or *N. benthamiana* or strain GV2260 for *N. benthamiana* (see Note 2).

2.2. Bacterial Growth and Transformation

1. Plates with LB solid medium supplemented with 50 mg/L kanamycin, 50 mg/L rifampicin, gentamycin 100 mg/L (see Note 3).
2. Tubes with LB liquid medium supplemented with 50 mg/L kanamycin, 50 mg/L rifampicin, gentamycin 100 mg/L.
3. Electroporator (e.g., BioRad Gene Pulser II).

2.3. Plant and Potting Materials

1. *P. somniferum* cv. Persian white seeds or *N. benthamiana* seeds (see Note 4).
2. Pots, pot trays, and clear covers.
3. Compost and vermiculite.
4. Fungicide (propamocarb) and/or GNAT OFF™ (Hydrogarden, UK) insecticide (see Note 5).
5. Plant fertilizer (e.g., MiracleGro™—Scott's Company Ltd., UK).
6. Light source (e.g., OSRAM™ L36 W/77 FLUORA light bulbs).

2.4. Plant Bacterial Inoculation

1. 3′–5′ Dimethoxy 4′-hydroxyacetophenone (acetosyringone) (200 mM stock solution in dimethyl formamide (DMF) or dimethylsulfoxide (DMSO)) (see Note 6).
2. 2-(*N*-morpholino) ethane sulfonic acid (MES).
3. $MgCl_2$ (1 M stock solution).
4. Silwet® L-77 surfactant (0.05%, v/v) (Lehle Seeds, USA).
5. VIGS buffer: 10 mM $MgCl^2$, 2 g/L MES and 200 μM acetosyringone in sterile water.

2.5. Materials for Leaf Infiltration or Vacuum Infiltration

1. Plastic needleless syringe (from 1 to 10 mL).
2. Razor blades.
3. Vacuum pump (see Note 10).

2.6. Control Materials

1. RNA isolation method (e.g., Trizol™, Invitrogen).
2. RNase-free DNase (e.g., Turbo DNA Free™, Invitrogen).
3. Reverse transcriptase (e.g., AMV reverse transcriptase, Promega) and oligo-dT primer.
4. Endogenous control primers (e.g., *ACTIN*).

3. Methods

3.1. Designing and Constructing pTRV2 Vectors (see Note 7)

1. Select a 300–700 bp (see Note 8) section of the target gene that needs to be silenced and clone into pTRV2 vector (preferably to the 3′ end of the gene of interest).
2. Introduce, separately, pTRV1 and pTRV2 plasmids into *A. tumefaciens* GV3101 or GV2260 by electroporation (25 μF and 2.5 kV for a 0.2 cm electrocuvette). Select transformants on LB agar plates containing kanamycin (50 mg/L), gentamycin (100 mg/L), and rifampicin (50 mg/L).
3. Screen colonies by polymerase chain reaction (PCR) or restriction digestion to confirm the presence of pTRV1 and pTRV2.

3.2. Preparing N. benthamiana Plantlets for Leaf Infiltration

1. Distribute approximately 50 seeds per large pot in a 1:1 compost:vermiculite mixture. Cover with Saran wrap. Place under continuous light regime at 25°C.
2. Seeds will germinate in approximately 3 days. Remove Saran wrap.
3. After two cotyledons have developed, the first leaf will emerge. Now transfer individual seedlings to smaller pots. Discard more slowly germinating seedlings so to have plantlets in the same developmental stage.
4. Grow plants until three true leaves have expanded on average.
5. For ease of infiltration, stop watering 2 days before infiltration so enough air is present in the expanded leaves. Infiltration can be performed as soon as three true leaves have expanded.

3.3. Preparing P. somniferum Seedlings for Vacuum Infiltration

1. Stratify *P. somniferum* seeds on vermiculite in the dark at 4°C for 7–10 days.
2. Transfer to a growth chamber for germination and growth 22°C with 20 h light (see Note 9).
3. Seeds should be visibly seen to germinate in 3–4 days after being transferred to the growth chamber.

3.4. Preparing Agrobacterium for Inoculation

1. Inoculate *Agrobacterium* containing plasmids pTRV1 and pTRV2 each into 5 mL of LB medium containing kanamycin (50 mg/L), gentamycin (100 mg/L), and rifampicin (50 mg/L). Grow the cultures overnight at 28°C.
2. Do the same for negative (empty vector) and positive control (e.g., *PHYTOENE DESATURASE*) constructs (see Note 1).
3. Inoculate the 5 mL overnight cultures individually into fresh 100 mL LB medium containing the same antibiotic selection and supplemented with 2 g/L MES and 200 μM acetosyringone. Grow the cultures overnight at 28°C.

4. Spin down the bacterial cultures at 4,000 × *g* for 15 min at 10°C.
5. Discard the supernatant and resuspend the pellets in VIGS buffer. Dilute with VIGS buffer to a final OD_{600} of 1.5.
6. Incubate the culture at room temperature overnight.
7. Mix *Agrobacterium* cultures containing the pTRV1 and pTRV2 with target gene in 1:1 ratio.

3.5. Leaf Infiltration of Agrobacterium into N. benthamiana

1. Use lab coat, gloves, and eye protection as infiltration could result in spraying of *Agrobacterium* mix.
2. Use the corner of a razor blade to wound the abaxial (under) epidermis of the leaves in two positions close to the petiole and two positions in the middle of the leaf.
3. For every small superficial wound, place a fingertip on the adaxial side of the leaf (upper) and position the syringe with TRV1-TRV2 mix to the wound.
4. Steadily infiltrate the leaf so that the medium spreads in the air of the leaf parenchyma.
5. Place pots in pot trays with clear cover to retain high humidity.

3.6. Vacuum Infiltration of Agrobacterium into P. somniferum (see Note 10)

1. Add 1 μL of Silwet for every 2 mL of *Agrobacterium* mix.
2. Remove the 4-week-old *P. somniferum* seedlings from vermiculite and rinse their roots with tap water.
3. Submerge the seedlings into the *Agrobacterium* mix.
4. Vacuum infiltrate for approximately 1 min.
5. Rinse seedlings in VIGS buffer and plant in pots containing a compost/vermiculite mix in a ratio of 1:1 presoaked in water containing some plant fertilizer (MiracleGro™) and insecticide (GNATOFF™).
6. Water planted seedlings regularly and maintain in growth chamber at 22°C with 20 h light cycle (see Note 4).
7. Silencing effects should be seen in approximately 14 days in new leaves depending on if the candidate gene is functional in vegetative or reproductive tissues.

3.7. Plant Growth and Phenotyping

1. Use the appropriate experimental design to grow plants for the developmental phenotype of interest. For instance, when measuring flowering time, randomize labeled control and VIGS plants.
2. Depending on the phenotype of interest, grow plants under strictly controlled conditions. For phenotypes in flower development, plants can be transferred to a greenhouse. For genes possibly

affecting flowering time, plants have to be kept in a growth chamber with controlled light, temperature, and humidity.

3. Screen for phenotypes. Always use the negative empty vector control for comparison. Use a positive control, such as the bleaching phenotype of *PDS* silencing to qualitatively control for silencing efficiency.

3.8. Controls for Silencing

1. Use (semi) quantitative RT-PCR to test for silencing using primers that preferably span the nested pTRV2 construct (see Note 11).
2. Compare plant tissues with a presumed phenotype to negative control plants. Collect biological replicates.
3. Isolate RNA from the tissues, use a DNase treatment, test for RNA integrity by gel electrophoresis, and quantify RNA concentration spectrophotometrically.
4. Reverse transcribe equal amounts of VIGS and control RNA to cDNA.
5. Optimize number of cycles in PCR for gene of interest and normalization gene such as *ACTIN*.
6. Use two to three technical replicates per biological replicate. Load PCR products to be compared on a single gel. Use the same UV and camera settings for every gel picture.
7. Measure band intensity using image software such as ImageJ.
8. Normalize gene of interest to endogenous control gene.

4. Notes

1. Empty TRV2 vector as a negative control and a gene that produces a visible phenotype when silenced such as *PHYTOENE DESATURASE* as a positive control. This should produce a bleached-leaf phenotype.
2. Our earlier work had shown that the *Agrobacterium* strain GV3101 was optimal in opium poppy (14).
3. pTRV vectors carry kanamycin resistance, and *Agrobacterium* strain GV3101 carries gentamycin and rifampicin resistance. The selection properties of the particular Agrobacteria varies and should be checked.
4. We use the *P. somniferum* cv. Persian white as it can grow well in small pots to a convenient stature of 1–2 ft. The accessions of *N. benthamiana* commonly used in research laboratories are probably all closely related. For more information, see ref. 11.

5. When seedlings are germinated, fungus development has to be avoided because fungus may attract fungus gnats, which feed from plant seedlings. This can be efficiently avoided by using a combination of insecticide (Gnat Off) with fungicide (propamocarb). Universal fertilizer can be supplemented to enhance growth of either *N. benthamiana* or *P. somniferum*.
6. DMSO can probably be used as an alternative solvent in cases where safety regulations do not allow the use of DMF.
7. It is advisable to have all constructs ready before starting to grow plants, as the time needed to prepare a construct can be difficult to predict.
8. We routinely use gene fragments of 280–350 bp for silencing constructs with good results. Analysis of the effects of fragment size has been conducted more systematically and shown that fragments of 300–1,300 bp are optimal (20). Also it is important to consider the effects of redundancy when designing a silencing fragment for a gene that is part of a larger family of related genes, i.e., avoid domains with high-sequence conservation, and perhaps use 3′ UTRs.
9. *P. somniferum* needs long day light cycles to flower so we use a cabinet set to 20 h of light or use a greenhouse in the spring/summer season as long as temperature can be maintained.
10. While we use vacuum infiltration to deliver the *Agrobacterium* into the poppy seedlings, this can vary depending on the species, and other delivery methods (that may be less labor-intensive) are possible. As described above in Subheading 3.5, leaf inoculation in *N. benthamiana* is efficient. The high density of latex in poppy leaves may account for this method not being effective in this species. Also be aware that if the vacuum pressure is too high, it can damage the young seedlings. For poppy seedlings, we use a maximum of 600–800 mb.
11. In our hands, real-time PCR is not a good method to test for silencing, even when primers are outside the construct. This is possibly because the amplicons in most real-time PCR experiments are very short (typically less than 100 bp to assure efficiency), and silencing might produce mRNA fragments of this size.

References

1. Ratcliff F, Martin-Hernandez AM, Baulcombe DC (2001) Tobacco rattle virus as a vector for analysis of gene function by silencing. Plant J 25:237–245
2. Liu Y, Nakayama N, Schiff M, Litt A, Irish VF, Dinesh-Kumar SP (2004) Virus induced gene silencing of a DEFICIENS ortholog in Nicotiana benthamiana. Plant Mol Biol 54:701–711
3. Senthil-Kumar M, Mysore KS (2011) Virus-induced gene silencing can persist for more than 2 years and also be transmitted to progeny seedlings in Nicotiana benthamiana and tomato. Plant Biotechnol J 9(7):797–806

4. Geuten K, Irish VF (2010) Hidden variability of floral homeotic B genes in Solanaceae provides a molecular basis for the evolution of novel functions. Plant Cell 22:2562–2578
5. Lu R, Martin-Hernandez AM, Peart JR, Malcuit I, Baulcombe DC (2003) Virus-induced gene silencing in plants. Methods 30:296–303
6. Yang SJ, Carter SA, Cole AB, Cheng NH, Nelson RS (2004) A natural variant of a host RNA-dependent RNA polymerase is associated with increased susceptibility to viruses by Nicotiana benthamiana. Proc Natl Acad Sci USA 101:6297–6302
7. Ying XB, Dong L, Zhu H, Duan CG, Du QS, Lv DQ, Fang YY, Garcia JA, Fang RX, Guo HS (2010) RNA-dependent RNA polymerase 1 from Nicotiana tabacum suppresses RNA silencing and enhances viral infection in Nicotiana benthamiana. Plant Cell 22:1358–1372
8. Baulcombe DC (1999) Fast forward genetics based on virus-induced gene silencing. Curr Opin Plant Biol 2:109–113
9. Mlotshwa S, Yang Z, Kim Y, Chen X (2006) Floral patterning defects induced by Arabidopsis APETALA2 and microRNA172 expression in Nicotiana benthamiana. Plant Mol Biol 61:781–793
10. Goodin MM, Zaitlin D, Naidu RA, Lommel SA (2008) Nicotiana benthamiana: its history and future as a model for plant-pathogen interactions. Mol Plant Microbe Interact 21:1015–1026
11. Dinesh-Kumar SP, Anandalakshmi R, Marathe R, Schiff M, Liu Y (2007) Virus-induced gene silencing. In: Grotewold E (ed) Plant functional genomics. Humana Press, Totowa
12. Hayward A, Padmanabhan M, Dinesh-Kumar SP (2011) Virus-induced gene silencing in nicotiana benthamiana and other plant species. Methods Mol Biol 678:55–63
13. Hileman LC, Drea S, Martino G, Litt A, Irish VF (2005) Virus-induced gene silencing is an effective tool for assaying gene function in the basal eudicot species Papaver somniferum (opium poppy). Plant J 44:334–341
14. Wege S, Scholz A, Gleissberg S, Becker A (2007) Highly efficient virus-induced gene silencing (VIGS) in California poppy (Eschscholzia californica): an evaluation of VIGS as a strategy to obtain functional data from non-model plants. Ann Bot 100:641–649
15. Gould B, Kramer EM (2007) Virus-induced gene silencing as a tool for functional analyses in the emerging model plant Aquilegia (columbine, Ranunculaceae). Plant Methods 3:6
16. Di Stilio VS, Kumar RA, Oddone AM, Tolkin TR, Salles P, McCarty K (2010) Virus-induced gene silencing as a tool for comparative functional studies in Thalictrum. PLoS One 5:e12064
17. Becker A, Lange M (2010) VIGS-genomics goes functional. Trends Plant Sci 15:1–4
18. Larkin PJ, Miller JA, Allen RS, Chitty JA, Gerlach WL, Frick S, Kutchan TM, Fist AJ (2007) Increasing morphinan alkaloid production by over-expressing codeinone reductase in transgenic Papaver somniferum. Plant Biotechnol J 5:26–37
19. Facchini PJ, De Luca V (2008) Opium poppy and Madagascar periwinkle: model non-model systems to investigate alkaloid biosynthesis in plants. Plant J 54:763–784
20. Liu E, Page JE (2008) Optimized cDNA libraries for virus-induced gene silencing (VIGS) using tobacco rattle virus. Plant Methods 4:5

Chapter 6

Virus-Induced Gene Silencing in the Rapid Cycling Columbine *Aquilegia coerulea* "Origami"

Bharti Sharma and Elena M. Kramer

Abstract

Aquilegia Origami is an emerging model system for ecology and evolution, which has numerous genetic and genomic tools. Virus-induced gene silencing (VIGS) has been established as an effective approach to study gene function in *Aquilegia*. In the current protocol, we demonstrate VIGS using *Agrobacterium* strain GV3101 carrying tobacco rattle virus (TRV)-based constructs to infect *Aquilegia coerulea* "Origami" plants via vacuum infiltration.

Key words: *Aquilegia,* VIGS, *Agrobacterium,* Tobacco rattle virus, Vacuum infiltration

1. Introduction

The genus *Aquilegia* (columbine) consists of ~70 species and is a member of the lower eudicot family Ranunculaceae (1). These species radiated rapidly in Eurasia and then expanded into North America over the past 1–3 myr (2, 3). *Aquilegia's* phylogenetic position between the established models of the core eudicots and monocot grasses, recent species diversification, and small genome size (~320–400 Mbp) make it an attractive system for a wide range of evolutionary, ecological, and developmental studies (4). In particular, the close evolutionary relationships among species of *Aquilegia* mean that they are highly interfertile and very similar at the level of DNA sequence (3, 5). The *Aquilegia* genome was recently sequenced by the Joint Genome Institute (JGI) of the US Dept. of Energy and is available in the Phytozome v7.0 database (6). Other genomic tools available for *Aquilegia* include an expressed sequence tag (EST) database (DFCI *Aquilegia* Gene Index) and two mapped

Annette Becker (ed.), *Virus-Induced Gene Silencing: Methods and Protocols*, Methods in Molecular Biology, vol. 975, DOI 10.1007/978-1-62703-278-0_6, © Springer Science+Business Media New York 2013

bacteria artificial chromosome (BAC) libraries (7, 8). *Aquilegia coerulea* "Origami," also called *Aquilegia x hybrida* "Origami," is a rapid cycling, herbaceous, and compact form of *Aquilegia* and was the basis for the JGI sequencing effort. The plants have a relatively short life cycle of ~4 months that can be completed in 4-in. pots with a short vernalization requirement of 2–3 weeks, which can be conducted at early developmental stages (8).

Although stable transformation is actively under development in *Aquilegia*, virus-induced gene silencing (VIGS), an RNAi-based reverse genetics technique, is currently our only tool for functional gene analysis. One of the real advantages of VIGS is its speed—phenotypes can be observed 2–3 weeks after infection. Previous studies using VIGS in *Aquilegia vulgaris* were carried out either by vacuum infiltration of seedlings or by wounding and injection of the basal rosette (9, 10). While the former can be used to obtain vegetative phenotypes (11), floral phenotypes in *A. vulgaris* require the less effective injection method (12). This is due to the fact that *A. vulgaris* has a very strong requirement for vernalization (8 weeks at 4°C) and is not receptive to floral induction until the 6–8 true leaf stage. Moreover, we have found that if *A. vulgaris* plants are VIGS treated as seedlings and then vernalized, silencing is abolished (E. Kramer, unpublished data). This challenge is effectively addressed through the use of *A. coerulea* "Origami" due to its minimal vernalization requirement and early competence to respond to floral induction. These plants can be vernalized at 4°C for 2–3 weeks at the 4–6 true leaf stage and then treated using vacuum infiltration. The efficiency of VIGS in "Origami" is relatively high, around 25–30%; however, it is important to note that there is a wide range of phenotypes. Strong silencing is observed in ~10% of treated plants (roughly a third of those showing silencing), while the remainder exhibits mild to moderate silenced phenotypes. Silenced sectors are common. This makes it critical to assess the degree of silencing using quantitative real-time PCR.

Here we describe both the growth and VIGS protocols for *A. coerulea* "Origami." Germination and initial growth can be completed in 3–4 weeks, followed by ~3 weeks of vernalization and immediate treatment for VIGS. Plants flower in another ~3 weeks, allowing rapid analysis of phenotypes. The compact size of "Origami" also facilitates growth and treatment of large numbers of plants, facilitating high-throughput approaches.

2. Materials

Prepare all solutions using ultrapure water (prepared by purifying deionized water to attain a sensitivity of 18 MΩ cm at 25°C) and analytical grade reagents. Prepare and store all reagents at room temperature (unless indicated otherwise).

1. pTRV1 and pTRV2 vectors and PCR primers that recognize these plasmids (see Note 1).
2. All components to carry out a PCR reaction and a PCR cycler.
3. T4 ligase and the corresponding buffer.
4. Restriction enzymes and the corresponding buffers.
5. Chemically competent *Escherichia coli* (e.g., Invitrogen One Shot Top10).
6. Electrocompetent *Agrobacterium tumefaciens* strain GV3101 (prepared as described in ref. 12) (see Note 2).
7. Standard LB liquid media and plates.
8. 60% glycerol, autoclaved or filter sterilized.
9. An electroporator, e.g., ECM399 Electroporator (BTX Genetronics, San Diego, CA).
10. PCR primers that add the appropriate restriction enzyme sites to the ends of your DNA fragment. We typically use *Xba* and *BamH1* sites in forward and reverse primers, respectively, to insert gene fragments of interest into TRV2 (see Note 3).
11. Filter-sterilized antibiotic stock solutions. pTRV1 and pTRV2 confer kanamycin resistance, and *Agrobacterium* GV3101 is resistant to gentamycin and rifampicin. Kanamycin and gentamycin 1,000× stock solutions should be prepared at 50 mg/ml in distilled water, while rifampicin 1,000× should be 25 mg/ml in methanol.
12. Prepare stock solution of 1 M 2-(*N*-morpholino)ethanesulfonic acid (MES) by dissolving 195.2 g/l in ultrapure water, followed by filter sterilization.
13. Prepare 0.1 M acetosyringone in methanol by dissolving 0.0196 g/ml (see Note 4).
14. Prepare 1 M $MgCl_2$ (hydrate) in ultrapure water by dissolving 0.2033 g/ml and filter sterilizing.
15. Sterilized glass 25 ml and 1,000 ml flasks (see Note 5).
16. Infiltration solution: 1 l = 10 ml 1 M MES, 10 ml 1 M $MgCl_2$, 2 ml 0.1 M acetosyringone, 978 ml ultrapure H_2O.
17. Silwet L-77.
18. Timer.
19. Vacuum source (see Note 6).

3. Methods

3.1. Growing *A. coerulea* "Origami"

1. Obtain *A. coerulea* "Origami" seeds from a reliable source, such as Swallowtail Seed Company or Goldsmith Seed Company (see Note 7). We typically use stocks of the "Red and White" variety (Fig. 1a, b).
2. Fill a 4×8 in. germination tray with moistened potting soil to a depth of ~4 in. and evenly distribute 50–100 seeds per tray. Seed does not need to be covered with soil.
3. Keep the trays covered with transparent plastic lids in order to maintain high humidity. Place into growth chambers at 18–20°C with 16 h days (full spectrum light). Check every other day to make sure soil remains moist.
4. Seeds will start germinating after 2 weeks. Let the seedlings reach two true leaves and then gently pull them out of soil and pot them individually into 4 in. pots. Take special care to water gently for the first few days. *A. coerulea* "Origami" will grow well in standard *Arabidopsis* growth chambers maintained at 18–20°C during the day/13–15°C at night with 16 h days (full spectrum light).
5. Let the seedlings grow until they have 4–6 true leaves (Fig. 1a), which generally takes 3–4 weeks. At this stage, they can be moved into vernalization at 4°C with short days (8 h days of full spectrum light) for 2–3 weeks.
6. After vernalization, remove plants back to original conditions (18–20°C during the day/13–15°C at night with 16 h days). Plants can be treated with VIGS within 1–2 days of removing from vernalization (see Subheading 3.3 below) and will start flowering 2–4 weeks after vernalization (see Note 8).

3.2. TRV2 Construct Preparation

1. PCR amplify a 200–500 bp fragment of your gene of interest using sequence-specific primers that add the appropriate restriction sites to the ends of the fragment.
2. After confirming that the PCR reaction worked specifically, column purify the product. Digest 1 μg each of the TRV2 vector and your gene fragment using the appropriate restriction enzymes. After ensuring that the digest has gone to completion, heat deactivate the restriction enzymes as described by the manufacturer. Column purify both the digested TRV2 vector and gene fragment (see Note 9).
3. The digested and cleaned vector and insert can be mixed in a 1:3 vector to insert ratio and ligated using T4 DNA ligase. The ligation can be carried out at room temperature for 10 min to 1 h.

Fig. 1. **(a)** *Aquilegia coerulea* "Origami" seedling with 4–6 true leaves. **(b)** *A. coerulea* "Origami Red and White" untreated flowers. **(c)** Experimental setup with desiccation chamber, vacuum pump, and cleaned seedlings. **(d)** Seedlings immersed in *Agrobacterium* infiltration medium. **(e)** Seedlings silenced for *AqPDS* showing distinctive photobleached phenotype. **(f–g)** *A. coerulea* "Origami Red and White" flowers silenced for *AqANS*. Note that silencing is not 100% in all organs, particularly in **(g)**.

4. Follow manufacturer's instructions to transform 4 μl of the ligation into chemically competent *E. coli* and then spread ~100 μl of transformed cells each onto two plates containing 50 mg/l kanamycin. Incubate plates at 37°C overnight.
5. PCR screen colonies using TRV2 primers 156F and 156R (see Note 1) in order to identify plasmids with intact inserts. It may be advisable to further PCR screen positive colonies using your original gene-specific primers in order to confirm that the insert in question is the desired fragment.
6. Grow 3 ml overnight cultures from the positive colonies and prepare plasmid DNA of the TRV2 construct. Test digest the plasmid DNA with your original restriction enzyme(s) to confirm correct sizes of vector and insert.
7. Gently mix 1 μl of the purified TRV2 plasmid into 50 μl of thawed GV3101 electrocompetent cells. Transfer the mixture to a chilled electroporation cuvette and shock at 2.4 kV for 5 ms (see Note 10). Add 1 ml of plain liquid LB to the cuvette and then transfer the mixture to a sterile culture tube. Shake for 2 h at 160 rpm at 28°C. Pellet the cells at 6,300–11,000 × *g* and resuspend in 200 μl of fresh LB by very gently vortexing. Spread cells onto 2–3 LB plates containing gentamycin 50 mg/l, kanamycin 50 mg/l, and rifampicin 25 mg/l. Incubate the plates at room temperature in the dark.
8. Likewise, transform GV3101 with the TRV1 plasmid.
9. Colonies will appear after ~48–72 h of incubation. Test colonies by PCR using appropriate PCR primers (see Note 1) and, for TRV2, subsequently with your gene-specific primers.
10. Start 3 ml overnight cultures from a positive colony, taking care not to let the culture grow more than 16 h. Next morning, prepare glycerol stocks for long-term storage by mixing 1 ml of *Agrobacterium* culture with 330 μl of sterile 60% glycerol in a cryo tube. Freeze tubes in liquid nitrogen for 10–20 s and then store them at −80°C.

3.3. Culture Preparation and Plant Infiltration

1. 3 days prior to the date you wish to infiltrate, streak out the TRV1 and TRV2 GV3101 cultures on appropriate antibiotic plates.
2. Select one positive colony from TRV1 and TRV2 and start 5 ml liquid cultures in LB with the appropriate antibiotics. Sterile 50 ml Falcon tubes are sufficient for these cultures.
3. Grow the cultures in a shaker at 28°C at 160–200 rpm for ~16 h.
4. The next day, use the entire 5 ml cultures to start 500 ml liquid LB cultures with the appropriate antibiotics for both TRV1 and TRV2. Use 1 l sterilized, baffled flasks. Add 100 μl freshly

prepared 0.1 M acetosyringone and 5 ml 1 M MES to the each flask containing 500 ml of culture.

5. Shake them overnight at 28°C at 160–200 rpm or until they reach an OD_{600} of 2.
6. Pellet the cultures by centrifugation for 15 min at 4,000 × *g* at 4°C.
7. Remove the supernatant and resuspend the pellet in infiltration buffer using gentle vortexing, targeting an OD_{600} of 2 (see Note 11).
8. Transfer the resuspended TRV1 and TRV2 solution into separate flasks and let them stand at room temperature for 3–4 h. Gently swirl the flasks every hour.
9. While the TRV cultures are resting, prepare the plants by removing them from their pots and gently washing the roots. This can be done by immersing them in a container filled with water and then by washing the roots under running tap water to remove any excess soil. Be careful not to break roots too much (see Note 12).
10. Once the plants are cleaned, mix an equal volume of the TRV1 and TRV2 cultures in a 1 l wide-mouthed beaker and add 0.005% Silwet L-77.
11. Completely immerse 5–10 plants in the mixed culture solution, place the beaker into a vacuum desiccation chamber, and apply strong vacuum to evacuate the chamber (Fig. 1c, d). Depending on the size of the chamber, this usually takes 15–30 s. Once the chamber is under strong vacuum, hold the pressure for 30 s–1 min. You should observe bubbles escaping from the seedlings immersed in the solution but the solution should not boil (see Notes 13 and 14).
12. Release the pressure by quickly opening the chamber to the outside atmosphere via the release valve. Remove the beaker from the chamber and carefully remove all of the seedlings. Place them in a paper-towel-lined tray to let the buffer drip completely from the roots (see Note 15).
13. Carefully repot the treated plants. Refilling the pots with fresh soil helps the roots of plants recover more robustly (see Note 15).
14. Place the plants back in the previous growth conditions and let them recover for 1 week before removing any dead foliage, which can be particularly susceptible to powdery mildew infection.

3.4. Characterizing Phenotypes

1. The plants will start showing silencing 2–3 weeks after infiltration (Fig. 1e–g). If you have used a secondary silencing marker (see Note 13), carefully examine plants for the expected phenotype. It is important to be aware that silencing may affect

entire organs/meristems or only sectors. Likewise, you may observe different degrees of target silencing and all the TRV2-borne genes may not be affected equivalently. For these reasons, it is critical to assess gene expression levels using quantitative RT-PCR.

2. For every plant showing silencing, collect whole organs separately, photodocumenting each one before preserving either in fixative for scanning electron microscopy (SEM) or in liquid nitrogen with long-term storage at –80°C for RNA extraction. Collect organs showing different phenotypes separately so that their degree of silencing can be correlated with the phenotypes. For floral phenotypes, we typically collect organs from each whorl separately, treating the stamen whorls as a single pool. Repeat this process for control samples.
3. To assess gene expression levels, start by preparing total RNA from control and experimental samples. Standard RNA preparation protocols or kits are suitable for *Aquilegia*. RNA from multiple control samples can be pooled, but we usually keep the silenced samples separate. Synthesize cDNA using standard reverse transcription enzymes (e.g., Invitrogen SuperScript II) or kits (see Note 16). Use standard quantitative real-time PCR to determine the relative expression levels of the gene of interest as well as any control targets, such as *ANS* or *PDS* (see Note 16). For silenced tissue, expression can be initially quantified in pooled cDNA samples, but it is useful to also quantify expression in individual organs with different phenotypes (see ref. 10 or 14 for examples).

4. Notes

1. We obtained the TRV1 and TRV2 plasmids from Prof. Dinesh-Kumar of Univ. CA, Davis (15). All of the experiments described here were conducted with the original multiple cloning site TRV2 (15), but the Dinesh-Kumar lab has also generated a Gateway-cloning version of TRV2 (15) as well as one for ligation-independent cloning (16). The most commonly used primers for TRV detection are the following: for TRV1, OYL195 5′-CTTGAAGAAGAAGACTTTCGAAGTCTC-3′ and OYL198 5′-GTAAAATCATTGATAACAACACAGACAAAC-3′; and for TRV2, 156F 5′-TTACTCAAGGAAGCACGATGAGC-3′ and 156R 5′-GAACCGTAGTTTAATGTCTTCGGG-3′.
2. We follow the Glazebrook and Weigel (13) protocol for preparing electrocompetent *Agrobacterium*. For best results, take special care to keep cultures cold between washes, and if

possible, carry out the entire preparation process in a 4°C room.

3. When designing your PCR primers, follow the manufacturer's guidelines for adding proper nucleotide overhangs before the selected restriction sites (e.g., some enzymes cannot cut a site with no flanking nucleotides). In addition, select a region of your target gene that does not contain highly conserved domains since this can lead to nonspecific silencing (unless that's what you want).
4. Acetosyringone solution should be made fresh no more than 24 h prior to the experiment.
5. We get the best results from baffled flasks.
6. We use a Vacuubrand RZ 9 vacuum pump that is capable of reaching 2×10^{-3} torr. This level of pressure is not essential; however, if your lab has strong in-house vacuum, you may be able to use that. The critical thing is that when vacuum is applied, you should see air bubbles rapidly coming out of the tissue.
7. Commercially available *A. coerulea* "Origami" is actually F1 seed derived from a cross of the Goldsmith Seed Company parental lines, one of which was the basis for DOE genome sequencing of *Aquilegia*. Unfortunately, this sequenced parental line appears to have a hypersensitive response to *Agrobacterium* infection that renders it unsuitable for VIGS. We have found that commercially available F1 seed, particularly that from Swallowtail Seeds, yields the best silencing results. We have noticed that some other seed sources are somewhat more variable in floral morphology, particularly petal number.
8. If you are only interested in vegetative phenotypes, plants can be VIGS treated at 3–4 leaves without vernalization.
9. The TRV2 plasmid is a quite large so use a column designed to clean 10–12 kb long fragments.
10. Store the sterile cuvettes at 4°C until ready to use.
11. The supernatant should be bleached and autoclaved prior to disposal. In our experience, achieving the correct OD is important to the success of the experiment, so dilute the solution with care.
12. Water the plants 1 or 2 days prior to the experiment since watering them the same day makes it hard to remove the soil from the roots.
13. Treat 50–150 plants with your experimental TRV2 construct and at least 30 plants with a control TRV2 construct, either the empty vector or a marker gene alone. We most commonly use silencing of *ANTHOCYANIDIN SYNTHASE* (*ANS*) as a

marker for both vegetative and floral silencing (Fig. 1f, g). *PHYTOENE DESATURASE* (*PDS*) can be used to mark vegetative tissue (Fig. 1e), but it can restrict normal leaf growth and so is not ideal.

14. Applying the vacuum for too long increases the mortality rate in small seedlings so do not exceed 2 min.
15. Clean the work area with 70% ethanol after the experiment is complete. All *Agrobacterium* supernatant and the remaining infiltration solution should be bleached or autoclaved before discarding.
16. If you want to test for presence of TRV1 and TRV2, it is necessary to make separate cDNA pools for each plasmid using the OYL198 and 156R primers, respectively, to synthesize the DNA. This cDNA can then be used with the appropriate TRV1 or TRV2 PCR primers.
17. We use the *Aquilegia* homolog of *ISOPENTYL PYROPHOSPHATE:DIMETHYLALLYL PYROPHOSPHATE ISOMERASE2 (AqIPP2)* to normalize qRT-PCR: Forward, 5′-CAGGTGAAGACGGACTGAAGTTAT; Reverse, 5′-CCAAGACTGGAAAAAAGACCACAC (17).

Acknowledgment

This work was supported by NSF award EF-0412727 to EMK.

References

1. Munz PA (1946) Aquilegia: the cultivated and wild columbines. Gentes Herb 7:1–150
2. Whittall JB, Voelckel C, Kliebenstein DJ et al (2006) Convergence, constraint and the role of gene expression during adaptive radiation: floral anthocyanins in Aquilegia. Mol Ecol 15:4645–4657
3. Hodges SA, Arnold ML (1994) Columbines: a geographically widespread species flock. Proc Natl Acad Sci USA 91:5129–5132
4. Kramer EM, Hodges SA (2010) Aquilegia as a model system for the evolution and ecology of petals. Philos Trans R Soc Lond B Biol Sci 365:477–490
5. Prazmo W (1965) Cytogenetic studies on the genus *Aquilegia*. IV. Fertility relationships among the *Aquilegia* species. Acta Soc Bot Pol 34:667–685
6. Phytozome v8.0.http://www.phytozome.net/search.php?method=Org_Acoerulea January 16th 2012
7. Fang GC, Blackmon BP, Henry DC et al (2010) Genomic tools development for Aquilegia: construction of a BAC-based physical map. BMC Genomics 11:621
8. Kramer EM (2009) Aquilegia: a new model for plant development, ecology, and evolution. Annu Rev Plant Biol 60:261–277
9. Gould B, Kramer EM (2007) Virus-induced gene silencing as a tool for functional analyses in the emerging model plant Aquilegia (columbine, Ranunculaceae). Plant Methods 3:6
10. Kramer EM, Holappa L, Gould B et al (2007) Elaboration of B gene function to include the identity of novel floral organs in the lower eudicot Aquilegia. Plant Cell 19:750–766
11. Blein T, Pulido A, Vialette-Guiraud A et al (2008) A conserved molecular framework for compound leaf development. Science 322: 1835–1839
12. Kramer EM, Holappa L, Gould B, Jaramillo MA, Setnikov D, Santiago P (2007) Elaboration

of B gene function to include the identity of novel floral organs in the lower eudicot Aquilegia (Ranunculaceae). Plant Cell 19: 750–766

13. Weigel D, Glazebrook J (2002) Arabidopsis: a laboratory manual. Cold Spring Harbor Press, Cold Spring Harbor, NY
14. Sharma B, Guo C, Kong H et al (2011) Petal-specific subfunctionalization of an APETALA3 paralog in the Ranunculales and its implications for petal evolution. New Phytol. doi: 10.1111/j.1469-8137.2011.03744.x
15. Liu YL, Schiff M, Dinesh-Kumar SP (2002) Virus-induced gene silencing in tomato. Plant J 31:777–786
16. Dong Y, Burch-Smith TM et al (2007) A ligation-independent cloning tobacco rattle virus vector for high-throughput virus-induced gene silencing identifies roles for NbMADS4-1 and -2 in floral development. Plant Phys 145:1161–1170
17. Ballerini E, Kramer EM (2011) The control of flowering time in the lower eudicot model Aquilegia. EvoDevo 2:4

Chapter 7

Virus-Induced Gene Silencing of the Alkaloid-Producing Basal Eudicot Model Plant *Eschscholzia californica* (California Poppy)

Dawit G. Tekleyohans, Sabrina Lange, and Annette Becker

Abstract

Eschscholzia californica (California poppy), a member of the basal eudicot family of the Papaveraceae, is an important species to study alkaloid biosynthesis and the effect of alkaloids on plant metabolism. More recently, it has also been developed as a model system to study the evolution of plant morphogenesis. While progress has been made towards establishing methods for generating genetically modified cell culture lines, transcriptome data and gene expression analysis, the stable transformation and subsequent regeneration of transgenic plants has proven extremely time consuming and difficult. Here, we describe in detail a method to transiently down regulate expression of a target gene by virus-induced gene silencing (VIGS) and the subsequent analysis of the VIGS treated plants. VIGS in *E. californica* allows for the study of gene function within 2 to 3 weeks after inoculation, and the method proves very efficient, enabling the rapid analysis of gene functions.

Key words: *Eschscholzia californica,* Tobacco rattle virus (TRV), pTRV, Virus-induced gene silencing, *Agrobacterium tumefaciens,* Gene function analysis, Quantitative real-time PCR, Histology

1. Introduction

Eschscholzia californica, basal eudicot, is a member of the Ranunculales order, which is the earliest branching eudicot lineage (1). *E. californica* belongs also to an early diverging lineage within the Ranunculales and even within the Papaveraceae *E. californica* holds a rather basal position (1, 2). As an early diverging eudicot, *E. californica* may serve as a model organism to bridge the morphological and molecular genetics gap between the distant groups of monocots and dicots which harbor the commonly studied plant models such as *Arabidopsis thaliana* (thale cress), *Nicotiana ssp.* (tobacco), *Medicago truncatula* (barrel medic), *Oryza sativa* (rice), or *Zea mays* (corn).

Annette Becker (ed.), *Virus-Induced Gene Silencing: Methods and Protocols*, Methods in Molecular Biology, vol. 975, DOI 10.1007/978-1-62703-278-0_7, © Springer Science+Business Media New York 2013

In addition to its phylogenetic position, *E. californica* is used as a model plant holding several advantages when compared to other basal eudicots: It is an annual or perennial plant with a generation time of 3 months when grown under constant light (3). It is diploid and has a relatively small genome size of around 1,115 Mb, which is roughly six times that of *A. thaliana* (4). 6,000 florally expressed sequence tags have been sequenced with high quality, and additional 311 Million bp have been gathered with Next Generation sequencing techniques, the *E. californica* microRNAs have been identified, and microarray chips for expression analysis are available (5–8).

E. californica is already a well-established model plant for the analysis of alkaloid biosynthesis and evolution of flower development. It is amenable to stable genetic transformation and regeneration mediated by *Agrobacterium tumefaciens* using cotyledons as explants (9) or seeds of a specific developmental stage (Yellina and Becker, personal communication). Floral and vegetative development has been described in detail, and protocols have been established for gene expression analysis by in situ hybridization (3, 10). Moreover, a BAC library covering the entire genome and a fast neutron irradiated mutant population exists (M. Lange, personal communication).

As an alkaloid producing organism, *E. californica*'s benzophenanthridine alkaloids biosynthesis, their specific activities and the plant's detoxification mechanisms have been a long standing focus of research (11–14).

Virus-induced gene silencing (VIGS) is a method that employs the plant's innate defense mechanism against viral infection. Once the virus has entered the plant cell it replicates and transcribe its DNA or RNA genome. Double-stranded RNA (dsRNA) intermediates produced during replication and/or transcription of the viral genome trigger the silencing mechanisms as they are recognized by the Dicer-like enzymes (15, 16). These bind to and cleave dsRNA molecules into short, approximately 21 nucleotide long short interfering RNAs (siRNAs). Single-stranded siRNAs are incorporated into the RNA-induced silencing complex (RISC) which screens for RNAs with sequences complementary to the siRNAs and degrades these RNAs (17). In the normal defense case, the RNAs targeted for degradation are only of viral origin, and thus the entire mechanism inhibits consistently viral RNA replication and transcription. Viral particles as well as siRNAs spread through the plant as systemic signals ensuring a rapid and efficient silencing of viral RNAs (18).

In the case of *E. californica*, a modified version of the bipartite genome of the tobacco rattle virus (TRV) is inserted into binary vectors and used for plant transformation. The modifications allow the insertion of a region of a gene of interest (GOI) into the viral RNA and enable the viral RNA to self-cleave from the transcribed

Table 1
Time line of VIGS experiments in *E. californica*

Activity	Timeframe	Remark
Growing *E. californica* plants	2 weeks	
Vector construction and verification	2 weeks	Confirm all clones by sequencing
Preparation of materials for plant infiltration	2 days	100 mL of each solution
Preparation of *Agrobacterium* cultures carrying the pTRV vectors for infiltration	4 days	
Infiltration of plants with *Agrobacterium* cells	1 day	50–100 plants per construct
Incubation of infiltrated plants	1 day	Incubate at 4°C
Observation of photobleaching on the control plants infiltrated with *pTRV2_EcPDS* construct	1–2 weeks after infiltration	Score the number of plants showing the photobleaching phenotype
First bud emergence	8 weeks after planting	Depending on the light condition

mRNA of the T-DNA (19). As the inserted partial sequence of the GOI is now part of the viral RNA and participates in all transcription and replication processes of the modified TRV, it will be recognized by Dicer-like enzymes and the RISC. Consequently, not only the viral RNA will be degraded but also the *E. californica* target gene. In theory, the transformation of a single plant cell with the entire bipartite modified TRV genome is sufficient for silencing in the entire plant.

In this chapter, we describe the method of VIGS in *E. californica* in detail. VIGS works extremely efficiently in *E. californica*, and a reduction of gene expression in the three first formed flowers is observed in 90–95% of all plants (10, 20) (Table 1).

2. Materials

Prepare all solutions (except for watering the plants) using ultrapure water (prepared by purifying deionized water to attain a sensitivity of 18.2 MΩ cm at 25°C) and analytical grade reagents. Prepare and store all reagents at room temperature (unless indicated otherwise). Diligently follow all waste disposal regulations when disposing waste materials.

2.1. Plant Material and Growth Conditions

E. californica seeds (Aurantiaca Orange King), purchased from B&T World Seeds, Aigues-Vives, France, can be stored at 4°C for several years until sowing. Three seeds are sown in moist soil (filling height about 8 cm) in one standard pot (9.5×9×9 cm—height, length, width) using a mixture of 70% soil (Einheitserde Typ 0, Wessels GmbH & Co. KG, Haren/Ems, Germany), 20% sand (TBG Lüssen, Germany) with 2 g/L Perlite (Wessels GmbH & Co. KG, Haren/Ems, Germany) and covered with approximately 1 cm of the soil mixture. Keep pots covered with plastic foil in the dark at 4°C for 3–5 days for stratification. After stratification remove plastic foil and keep the plants in a greenhouse under long day condition with at least 16 h light provided by mercury lamps and a temperature between 20 and 22°C in the day time and 16–18°C at night. Water pots daily with tap water and let the plants grow for 3 weeks until they develop two to three foliage leaves. For further VIGS-inoculation keep just two healthy plants per pot and avoid replanting of plantlets as it will damage the sensitive root system. Prepare for at least 50 plants per gene of interest.

2.2. Vectors Construction and A. tumefaciens Strain for Inoculation

2.2.1. Vector System (19)

The vector system to be used consists of the bipartite TRV genome distributed in two vectors, pTRV2 and pTRV1. The pTRV1 vector contains genes coding for movement protein, replicase proteins and cysteine-rich protein. The pTRV2 vector includes genes encoding a coat protein and two non-structural proteins. Both parts of the viral genome are driven by a duplicated CaMV 35S promoter and terminated with nopaline synthase. The pYL44 and the pCAMBIA0390 vector backbone were used for the construction of pTRV1 and pTRV2 vectors respectively. Both binary vectors are resistant to the antibiotic kanamycin, and the pTRV2 vector also contains a multiple cloning site (MCS) where the gene of interest can be inserted.

2.2.2. Materials for pTRV2 Construction Containing the GOI

1. Plasmid pTRV2.
   (a) The MCS contains the following restriction sites: *Eco*RI, *Xba*I, *Nco*I, *Bam*HI, *Kpn*I, *Xma*I, *Sac*I, *Xho*I, *Xma*I, *Sma*I.
2. Gene-specific primers including restriction sites:
   Design primers which can PCR amplify a unique region within the gene of interest, a fragment containing 3′UTR is desirable. The optimum size of the amplicon should be in the range of 400–500 bp. The primers should contain a restriction site at their 5′ end which is compatible with the restriction sites present in the multiple cloning site of the pTRV2 vector. We generally use *Bam*HI, *Eco*RI, and/or *Xba*I. If the GOI is a member of a gene family, the amplicon should not contain a conserved region shared by the other members when aiming at gene-specific silencing.

3. Polymerase chain reaction (PCR) components: Reaction buffer, $MgCl_2$ (if not included in the reaction buffer), ddH_2O, Forward and Reverse Primers, Thermocycler (e.g., Eppendorf, Hamburg, Germany), High Fidelity proof reading polymerase (Fermentas, Sankt Leon-Rot, Germany), dNTP mix, and cDNA as template.
4. Restriction enzymes and corresponding buffers matching the restriction sites present in your primers and the pTRV2 multiple cloning site.
5. 50× TAE buffer: add 121 g Tris base and dissolve it in 250 mL ddH_2O. Then add 28.6 mL acetic acid and 50 mL 0.5 M EDTA pH 8.0. Adjust the volume to 500 mL with ddH_2O.
6. 1% Agarose Gel: Weigh 5 g Agarose (Biozym, Hessisch Oldendorf, Germany) in a 500 mL glass bottle containing 500 mL 1× TAE buffer. Heat to dissolve the agarose in a microwave and finally add 0.005% SERVA DNA Stain G (SERVA Electrophoresis GmBH, Heidelberg, Germany).
7. Gel electrophoresis apparatus (VWR, Hannover, Germany).
8. PCR product purification and/or gel extraction kit (e.g., Macherey-Nagel GmbH & Co. KG, Düren, Germany).
9. NanoDrop spectrophotometer (Thermo Scientific, Wilmington, USA).
10. T4 Ligase plus its corresponding buffer (New England Biolabs GmbH, Frankfurt am Main, Germany).
11. JM109 competent *E. coli* strain (Promega, Mannheim, Germany).
12. Water Bath.
13. SOC media: Weigh 20 g tryptone, 5 g yeast extract, and 0.5 g NaCl in a 1,000 mL glass bottle. Add 950 mL deionized water and stir until it is dissolved. Add 10 mL of a 250 mM solution KCl (dissolve 1.89 g KCl in 100 mL deionized water). Adjust pH to 7.0 with 5 M NaOH (~200 μL) and add up to 1,000 mL with deionized water, then sterilize. After media is cooled down to 60°C, add 5 mL of a sterile 2 M solution of $MgCl_2$ (dissolve 19 g of $MgCl_2$ in 100 mL of deionized water and sterilize). Next add 20 mL of a sterile 1 M solution of glucose (dissolve 18 g of glucose in 100 mL deionized water and pass it through a 0.20 μm filter) (see Note 1).
14. LB liquid and solid medium containing kanamycin: Weigh 10 g tryptone, 5 g yeast extract, and 5 g NaCl in a 1,000 mL glass bottle. Add 950 mL deionized water and stir until dissolved. Adjust pH to 7.0 with 5 M NaOH (~200 μL) and adjust volume to 1,000 mL with deionized water. For solid medium add 15 g agar and then sterilize. Once LB is cooled

down to 60°C, add sterile filtered kanamycin sulfate at a concentration of 100 μg/mL.

15. Kits for *E. coli* plasmid isolation (e.g., Macherey-Nagel GmbH & Co. KG, Düren, Germany).

2.2.3. A. tumefaciens Strains

1. *A. tumefaciens* strain GV3101 (21) is used for infiltration of plants. In order to select transformed Agrobacterium cells, the antibiotics kanamycin sulfate and gentamicin sulfate must be used at a working concentration of 100 and 50 μg/mL, respectively.
2. Vector pTRV1 containing *A. tumefaciens* strain.
3. Vector pTRV2-*EcPDS* containing *A. tumefaciens* strain. This, combined with pTRV1, is used as positive control in our lab to silence the phytoene desaturase gene causing visible photobleaching of leaves and colorless floral organs (16).

2.3. Solutions and Equipment for Preparing Electrocompetent A. tumefaciens Strain GV3101 Cells

1. *A. tumefaciens* strain GV3101.
2. LB medium containing gentamicin: Weigh 10 g tryptone, 5 g yeast extract, and 5 g NaCl in a 1,000 mL glass bottle. Add 950 mL deionized water and stir until dissolved. Adjust pH to 7.0 with 5 M NaOH (~200 μL) and prepare a total volume of 1,000 mL with deionized water, then sterilize. Once LB is cooled down to 60°C, add sterile filtered gentamicin.
3. 10% (v/v) sterile glycerol.
4. High speed centrifuge (e.g., Eppendorf, Hamburg, Germany).
5. Shaker.
6. Liquid N_2.

2.4. Materials Needed for Transformation of A. tumefaciens Strain GV 3101

1. Bio-Rad Electroporation Apparatus—MicroPulser (Bio-Rad, München, Germany).
2. Electroporation cuvettes—2 mm electroporation gap (Biozym, Hess. Oldendorf, Germany).
3. Electrocompetent *A. tumefaciens* GV3101.
4. SOC medium (see Chapter 2.2).
5. LB medium gentamicin (see Chapter 2.2 for LB medium preparation). Once LB is cooled down to 60°C, add sterile filtered gentamicin sulfate at a concentration of 50 μg/mL.

2.5. Material for Plant Infiltration

2.5.1. Injection Solutions

1. 10× $MgCl_2$ (100 mM): Weigh 2.03 g $MgCl_2$ and add 100 mL water. Stir until is dissolved.
2. 10× 2-(*N*-morpholino) ethanesulfonic acid (MES) (1 mM): 19 mg MES in 100 mL water.

3. 10× Acetosyringone (100 mM): 1.95 g acetosyringone in 100 mL absolute ethanol.
   Filter all solutions into sterile 50 mL Falcon™ tubes using a 0.20 μm sterile, non-pyrogenic filter (CA-Membrane, Sartorius, Germany) and a sterile 20 mL syringe with Luer-Lock adapter. For storage of the solutions, see Note 2.
4. Injection buffer:
   For 20 mL of injection buffer mix 2 mL of each injection solution 1, 2, and 3 and then add 14 mL of water and use directly for injection. Always prepare this buffer fresh.
5. 0.45 × 25 mm needle attached to a 2 mL syringe.

2.6. Material for Gene Expression Analysis

1. RNA extraction for 10–100 mg plant material: Use the plant-rna-OLS® Kit (Omni Life Science, Germany) according to the manufacturer instructions. RNA extraction for 0.1–10 mg plant material: Use the RNeasy® Micro Kit (Qiagen, Hilden, Germany) according to the manufacturer instructions (see Note 3).
2. RNA-gel components:
   10× (N-morpholino)propanesulfonic acid (MOPS): Weigh 41.8 g of MOPS (200 mM), 4.1 g sodium acetate and 3.7 g Ethylenediaminetetraacetic acid (EDTA) and add 900 mL Milli-Q H2O. Stir until everything is dissolved and adjust pH 7.0 with 1 M NaOH. Make up the volume to 1 L with water (see Note 4).
3. RNA sample buffer: Mix 70 μL formamide (deionized), 27 μL formaldehyde (35%), and 3 μL 10× 3-MOPS (see Note 5).
4. RNA loading buffer: Add 2.5 g Ficoll type 400, 2 mL of 0.5 M EDTA solution (pH 8.0), 25% (w/v) bromophenol blue and adjust the volume to 10 mL with water (see Note 6).
5. 1% aqueous ethidium bromide solution.
6. Reverse transcription reaction components: The RevertAid™ H Minus First Strand cDNA Synthesis Kit (Fermentas, Sankt Leon-Rot, Germany) is used according to the manufacturer's instruction (see Note 7).
7. qRT-PCR components: qRT-PCR is carried out with the LightCycler® 480 using the SYBR Green I Master kit (Roche, Mannheim, Germany) and 0.8–1.2 pM target gene primers (see Note 8).

2.7. Materials, Solutions and Buffers for Histological Analyses

1. Fixation Solution (FAE): Mix 6 mL formaldehyde (37%), 10 mL acetic acid, 120 mL absolute ethanol, 64 mL of deionized water, and one or two drops of Tween 20 in a clean 250 mL bottle.

2. Dehydration and staining solutions: Prepare the following components separately:
   - 30% ethanol: mix 30 mL absolute ethanol with 70 mL water.
   - 50% ethanol: mix 50 mL absolute ethanol with 50 mL water.
   - 70% ethanol: mix 70 mL absolute ethanol with 30 mL water.
   - 85% ethanol: mix 85 mL absolute ethanol with 15 mL water.
   - 95% ethanol: mix 95 mL absolute ethanol with 5 mL water.
   - 0.1% Safranin-O: Dissolve 100 mg Safranin-O (Carl Roth, Karlsruhe, Germany) in 99 mL of 30% ethanol.
   - 1% (w/v) Sodium bicarbonate: Weigh 1 g of Sodium bicarbonate and dissolve it in 99 mL of 30% ethanol.
   - 0.2% (w/v) Fast Green Solution: Weigh 200 mg of Fast Green (Carl Roth, Germany) and dissolve it in 99 mL of absolute ethanol.
3. Paraplast Plus (Carl Roth, Karlsruhe, Germany).
4. Glass slides "Superfrost Plus" and cover slips for microscopy (Thermo Scientific/Menzel GmbH, Braunschweig, Germany).
5. Heating plate (Medite, Burgdorf, Germany).
6. Rotary microtome, e.g., Hyrax M25 (Carl Zeiss, Jena, Germany).
7. Entellan (Merck Chemicals, Darmstadt, Germany).
8. Incubator, e.g., GFL 7601 (GFL, Burgwedel, Germany).
9. 100% Ethanol.
10. 1% Sodium bicarbonate (in 30% ethanol): Weigh 1 g Sodium bicarbonate and dissolve it in 100 mL 30% ethanol.

3. Methods

3.1. Construction of the pTRV2 Vector Containing a Fragment of the Gene of Interest

1. Amplify the gene of interest using PCR thermocycler and run 5 μL of the product on 1% Agarose Gel for 45 min at 100 V.
2. Elute the DNA fragment from the gel using DNA elution kit.
3. Measure the gel eluted DNA concentration and its quality on NanoDrop.
4. Restrict digest 0.5–1 μg of the eluted DNA and pTRV2 vector with appropriate restriction enzyme.
5. Load the restriction digest on a 1% agarose gel and run it for 45 min at 100 V.
6. Repeat steps 2 and 3.

7. Use a 5:1 insert to vector ratio for the ligation protocol. Calculate the amount of insert DNA required for a 50 ng of vector using the following formula.

$$\text{ng of insert required} = \frac{(50\text{ng of vector})\,(\text{insert size in bp})}{\text{Vector size in bp}} \times 5$$

8. Set up a ligation mix following the manufacturer's instructions using T4 ligase.
9. Set the water bath at 42°C and keep SOC medium in it.
10. Add 10 μL of the ligation product into 50 μL of chemical competent *E. coli* (JM109) cells and incubate on ice for 10 min.
11. Transfer the cell suspension into a preheated water bath for 150 s at 42°C.
12. Place the cell suspension on ice for 2 min and add 1 mL of prewarmed SOC medium.
13. Keep the cells shaking at 37°C for 1 h and then plate them on kanamycin containing solid LB medium.
14. On the next day count the number colonies grown per plate, analyze the presence of the plasmid by doing colony PCR (see Note 9), and analyze the PCR product on 1% agarose gel.
15. Prepare liquid overnight culture from the positive colonies by growing them in kanamycin containing 5 mL liquid LB medium. Incubate while shaking the culture overnight at 37°C.
16. Isolate the plasmid from the overnight culture following the manufacturer protocol.
17. Sequence the plasmid isolated to verify the correct insert (see Note 10).

3.2. Preparation of Electrocompetent A. tumefaciens (22, 23)

1. Inoculate 500 mL LB media containing gentamicin sulfate at a concentration of 50 μg/mL with 5 mL of a fresh saturated culture of *A. tumefaciens* and incubate overnight at 28°C with shaking at ~130 rpm.
2. Once the stationary phase with $OD_{550} = 0.8$ is reached, take 150 mL of the culture and inoculate 350 mL fresh LB media.
3. Shake for 2–3 h until $OD_{550} = 0.8$ is reached.
4. Chill suspension on ice for 20 min and collect cells by centrifugation at 4,000 × *g* for 10 min at 4°C (see Note 11).
5. Discard supernatant and resuspend cells in 10 mL of ice-cold water until the pellet is completely dissolved. Adjust the suspension volume to 250 mL with ice-cold sterile double distilled water.

6. Repeat steps 4 and 5 twice, resuspending the cells in the final volume of 250 mL and 50 mL of ice-cold sterile double distilled water respectively.
7. Pellet cells as in step 4 and resuspend them in 5 mL of 10% sterile and ice-cold glycerol.
8. Dispense the cells in 50 μL aliquots and immediately snap-freeze in liquid nitrogen (see Note 12).

3.3. *A. tumefaciens* Transformation

1. Thaw an aliquot of electrocompetent cells on ice.
2. Add 1 μL of ligation reaction (or plasmid) and incubate for 1 min on ice.
3. Transfer cell mix to a precooled electroporation cuvette, insert into MicroPulser and electro-pulse at 2.5 kV for 5.5 ms.
4. Immediately add 1 mL of SOC media (prewarmed at 28°C) and keep suspension shaking at 28°C for 1 h.
5. Plate culture on selective LB-agar plates in different volumes, e.g., 10, 20, and 50 μL. Incubate for 2–3 days at 28°C until colonies are visible.

3.4. Infiltration of *E. californica*

Water the plants in the morning on the day of infiltration with the *A. tumefaciens* strains. Label each pot properly reference to the applied VIGS-construct, e.g., pTRV2-*EcPDS*.

3.4.1. Infiltration of Plants with A. tumefaciens for VIGS (10)

Day 1. Prepare 4 mL LB cultures in sterile test tubes containing kanamycin and gentamicin and inoculate with a single colony or 2 μL of glycerol stock. Incubate overnight at 28°C and shake at approximately 130–150 rpm. Prepare a separate culture for each pTRV2 construct. Use an empty vector control (pTRV2_Empty and pTRV1) and a positive control (pTRV2-*EcPDS* and pTRV1).

Day 2. Inoculate 40 mL of LB + kanamycin and gentamicin using the 4 mL well-grown (OD_{550} 0.8–1.2) overnight culture. Prepare again 40 mL for each culture in sterile 100 mL Erlenmeyer flasks. Incubate the strains for approx. 3–4 h while shaking at 28°C until the cultures reach an OD_{550} 0.6–0.8.

3.4.2. Preparation for Injection

Spin down cultures in 50 mL Falcon™ tubes at 2,862 × *g* for 20 min at RT. Discard supernatant and dissolve each pellet in 20 mL of injection buffer by vortexing thoroughly. Mix each pTRV2-x carrying culture with a pTRV1 culture (1:1) by inversion to obtain a homogeneous suspension.

3.4.3. Injection of Plants

Inject each plant with 100–150 μL of *Agrobacterium* suspension mix by inserting the needle attached to a syringe carefully vertically into the apical most region of the plant (Fig. 1). Take care not to destroy the shoot apical meristem (SAM). Keep plants covered with plastic foil overnight at 8–10°C in the dark and transfer to

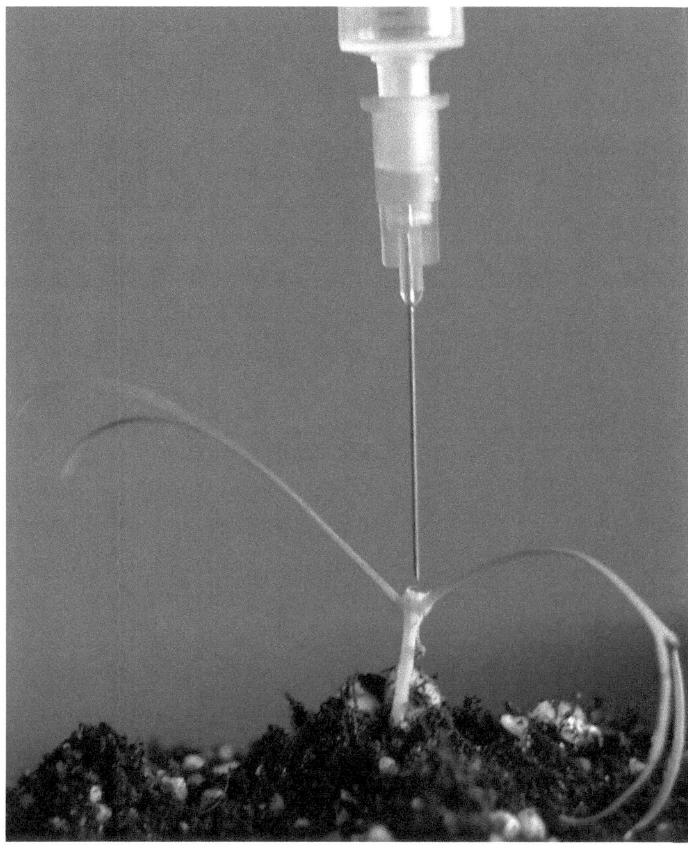

Fig. 1. Infiltration of *E. californica* in the four to five leaf stage with *Agrobacterium tumefaciens* carrying VIGS vectors. The needle for Agro-injection must target the shoot apical meristem for efficient infiltration.

greenhouse the next day. Remove covers and continue to water plants regularly with tap water. Photobleaching of the plants inoculated with pTRV1 and pTRV2-*EcPDS* can be observed after 12–14 days and the plants will flower in about 6–10 weeks.

3.5. RNA Isolation

In order to quantify the degree of gene silencing in VIGS treated plants, RNA sample should be taken carefully from both the treated and untreated plants. For genes which are expressed in the floral organs, it is recommended to harvest the first flower for RNA isolation since the effect of VIGS is strongest in the first formed flowers.

3.6. RNA Gel Electrophoreses

3.6.1. Formaldehyde Agarose Gel

1. Prepare a gel by mixing 36.1 mL Milli-Q H_2O, 5 mL 10× MOPS, and 0.75 g standard agarose in a 100 mL Erlenmeyer flask. Heat carefully in a microwave until the agarose is dissolved completely.

2. Once dissolved, shake the flask by hand. Always use heat protective gloves and avoid producing bubbles by shaking too vigorously. All work from now onwards should be performed under a clean chemical hood.
3. Cool down the mixture to approx. 60°C and add 8.9 mL of formaldehyde (35%).
4. Cast the gel immediately and allow it to solidify.
5. Place the gel in a horizontal electrophoresis apparatus and add 1× MOPS buffer to the reservoir until the buffer covers the agarose gel at least 1 cm high.

3.6.2. RNA-Samples

1. Mix 1 μg of RNA with 15 μL fresh sample buffer and 2 μL RNA loading buffer.
2. Denature the RNA by incubating the mix for 10 min at 60°C, and then place it immediately on ice.
3. Add 0.5 μL ethidium bromide (1% solution) to each sample. Wear nitrile gloves while working with ethidium bromide and discard contaminated disposals in special containers according to waste treatments of the biosafety data sheet.
4. Load samples into wells and start running the gel at 85 V for 45 min.
5. Visualize the RNA fragments on a long wave UV light box.

3.7. Reverse Transcription of RNA into cDNA

0.5–1 μg of total RNA is reverse transcribed into cDNA using random hexamer primers and following the manufacturer protocol. Store undiluted cDNA samples at −80°C.

3.8. Expression Analysis by qRT-PCR

Quantitative Real-Time PCR is carried out to assess the ability of the VIGS treatment in reduction of the target gene expression. As an internal controls, we recommend the primer pair Actin_QRT_Fw (5′-GGAATAGTGAGCAACTGGGA-3′) and Actin_QRT_Rev (5′-CAGCACAATACCTGTAGTACGA-3′) for the gene *ACTIN* and the primer pair GAPDH_QRT_Fw (5′-GCTTCCTTCAA CATCATTCC-3′) and GAPDH_QRT_Rev (5′-AGTTGCCTTCT TCTCAAGTC-3′) for the gene *GAPDH* (20). Before starting the qRT-PCR optimize the primer set working concentration and evaluate the efficiency of your target gene primers with a primer efficiency test. Generally, try to adhere as closely as possible to the MIQE guidelines for qRT-PCR (24). Produce three technical replicas for each cDNA pool and analyze an appropriate number of VIGS treated plants to obtain statistically sound data. Compare the gene expression of plants treated with the target gene VIGS construct with untreated plants and plants treated with the empty pTRV2 vector.

In our lab, qRT-PCR PCR is performed in a qPCR cycler (e.g., LightCycler® 480 (Roche, Germany)) with the following cycling conditions: initial heating of 95°C for 5 min, 45 cycles of 10 s at 95°C, 10 s at primer annealing temperature, and 10 s at 72°C (20). C_t values and normalization is carried out by LightCycler® 480 Software 1.5.0. Analysis of the qRT-PCR data is carried out as described in (25).

3.9. Assessing Morphological Changes Resulting from VIGS Treatment by Histological Analysis (10)

Gene silencing of target genes may result in morphological changes of the treated plants. For the histological analysis, the tissue to be observed needs to be fixed first and then embedded into paraffin-like wax, sectioned on a microtome, dewaxed, and finally stained.

3.9.1. Fixation of the Tissue

1. Place plant tissue into freshly made FAE solution until it is completely covered (see Notes 13, 14).
2. Apply vacuum (~4×10^{-4} mbar) for 10 min to degas tissue. The tissue should sink to the bottom of the tube.
3. Change FAE carefully to not expose the tissue to air and incubate for 24 h at 4°C. If air bubbles continue to reside in the tissue apply vacuum again for 10–20 min.
4. Wash tissue 2×5 min in 70% EtOH (p.a.) on ice and store fixed tissue in 70% EtOH at −20°C (see Note 15).

3.9.2. Dehydration of the Fixed Tissue

1. Dehydration of the tissue is achieved by passing it through an ethanol series. Remove the 70% EtOH by careful pipetting while taking care that the tissue is not exposed to air. All following steps are carried out at room temperature.
2. Replace 70% EtOH with 85% EtOH and incubate for 90 min.
3. Remove the 85% EtOH and replace with 90% EtOH, incubate for 90 min.
4. Transfer the tissue to 100% EtOH, incubate for 90 min, and repeat the last step once. If the tissue does not remain at the bottom of the tube, degas again by applying vacuum for 10–20 min.

3.9.3. Paraplast Embedding

1. Transfer the tissue into the following solution and incubate it in EtOH/Rotihistol mix series in the order from 3:1, 2:1, 1:1, and 1:2 ratios at room temperature, incubate for 60 min each.
2. Transfer the tissue to 100% Rotihistol and incubate at room temperature for 60 min, then transfer to 100% fresh Rotihistol and incubate as before.
3. Prepare a bed of solid paraplast beads in a small beaker and then remove most of the Rotihistol from your tissue sample without exposing it to the air and pour it on top of the solid paraplast bed (see Note 16).
4. Leave it overnight at 58–60°C to melt the paraplast.

5. Replace with fresh, molten paraplast and incubate at 60°C, ensuring that the temperature remains below 62°C, as the DMSO in the paraplast is temperature sensitive.
6. Exchange paraplast twice per day for 3–5 days.
7. Transfer tissue pieces to freshly molten paraplast. Preheat a heating block to 58–60°C and place a petri dish on it. Pour your samples into the petri dish and orientate your tissue as preferred. Cool down at RT and store at RT, or for longer storage at 4°C.

3.9.4. Sectioning

Section the tissue on a microtome into slices of 5–7 μm thickness. Place the paraplast slices on a glass slide and add double distilled water until the slices float and stretch out on the water surface. Remove the water and place slides on heating plate at 42°C for overnight and then store slides at 4°C until use.

3.9.5. Staining with Safranin O and Alcoholic Fast Green

1. Place slides in the slide holder and put them in the glass dish. Incubate twice for 10 min in Rotihistol followed by 2 min incubation in 100% Ethanol.
2. Incubate the slides in an ethanol series 95, 85, 70, 50, and 30% for 5 min in each of them.
3. Stain slides in 0,1% Safranin O-solution (made with 30% EtOH) for 30 min and then rinse it with 30% Ethanol and place slides in 1% Sodium bicarbonate (in 30% EtOH) for 5 min.
4. Rinse the slide in 30% and 50% Ethanol separately for 5 min in each.
5. Incubate slides for 60–90 s in 0.2% alcoholic Fast Green solution (in 100% EtOH) and wash it twice for 5 min with 100% EtOH and lastly incubate it in Rotihistol for 5 min.
6. Dry slides and mount sections with a cover slip and mount in Entellan.

The stained section can be observed with a standard research microscope to describe possible morphological changes based on VIGS-mediated down regulation of gene expression.

4. Notes

1. Avoid pulling the plunger while filter is connected to syringe, as low-pressure can damage the filter.
2. One set of injection solutions should be kept at 4°C in the dark and is stable up to 6 months. The second set can be stored at −20°C for long term.

3. Store RNA at −80°C. Keep RNA on ice constantly while proceeding with, e.g., electrophoresis, reverse transcription.
4. Store at 4°C in the dark for 1 month maximum—it yellows with age if exposed to light and should not be used once it has a dark yellow color. Do not autoclave!
5. Work under a fume hood. Always prepare this buffer freshly and dispose in special containers according to waste treatments of safety data sheet after use.
6. Loading buffer can be stored long term at 4°C in the dark.
7. Primer choice for the cDNA synthesis depends on the type of gene targeted.
8. Primers have to be analyzed for their suitability in a separate primer concentration and efficiency test. *Eschscholzia Actin2* and *GAPDH* are used as reference genes in our lab.
9. Pick colonies from the plate and resuspend them in 10 μL ddH_2O. Use 2 μL from these as a template for the colony PCR.
10. Sequence more than one clone and analyze the sequence data properly before proceeding to next step.
11. From this step on always keep cells on ice.
12. Store electrocompetent cells at −80°C until use.
13. When fixing flower buds, remove the bud tips (enhances penetration of fixative).
14. All the steps which involve formaldehyde should be done under a fume hood due to its toxicity.
15. Once the tissue is fixed, it can be stored in 70% EtOH for several months at −20°C.
16. Paraplast application should be performed quickly to avoid its solidification.

Acknowledgments

The *E. californica* work in A.B.'s lab is largely funded by the German Research Foundation (DFG grants BE 2547/6-1, 6-2, 7-2, 8-1).

References

1. Wang W, Lu A-M, Ren Y et al (2009) Phylogeny and classification of Ranunculales: evidence from four molecular loci and morphological data. Perspect Plant Ecol Evol Syst 11(2):81–110
2. Kadereit JW, Blattner FR, Jork KB et al (1995) The phylogeny of the Papaveraceae s.l.: morphological, geographical and ecological implications. Plant Syst Evol 9:133–145

3. Becker A, Gleissberg S, Smyth DR (2005) Floral and vegetative morphogenesis in California poppy (*Eschscholzia californica* Cham). Int J Plant Sci 166(4):537–555
4. Bennett MD, Smith JB (1976) Nuclear DNA amounts in angiosperms. Philos Trans R Soc Lond B Biol Sci 274(933):227–274
5. Barakat A, Wall K, Leebens-Mack J et al (2007) Large-scale identification of microRNAs from a basal eudicot (*Eschscholzia californica*) and conservation in flowering plants. Plant J 51(6):991–1003
6. Carlson J, Leebens-Mack JH, Wall PK et al (2006) EST database for early flower development in California poppy (*Eschscholzia californica* Cham., Papaveraceae) tags over 6000 genes from a basal eudicot. Plant Mol Biol 62(3):351–369
7. Wall PK, Leebens-Mack J, Chanderbali AS et al (2009) Comparison of next generation sequencing technologies for transcriptome characterization. BMC Genomics 10:347
8. Zahn LM, Ma X, Altman NS et al (2010) Comparative transcriptomics among floral organs of the basal eudicot Eschscholzia californica as reference for floral evolutionary developmental studies. Genome Biol 11(10):R101
9. Park SU, Facchini PJ (2000) Agrobacterium-mediated genetic transformation of California poppy, *Eschscholzia californica* Cham., via somatic embryogenesis. Plant Cell Rep 19(10):1006–1012
10. Orashakova S, Lange M, Lange S et al (2009) The *CRABS CLAW* ortholog from California poppy (*Eschscholzia californica*, Papaveraceae), *EcCRC*, is involved in floral meristem termination, gynoecium differentiation and ovule initiation. Plant J 58(4):682–693
11. Angelova S, Buchheim M, Frowitter D et al (2010) Overproduction of alkaloid phytoalexins in California poppy cells is associated with the co-expression of biosynthetic and stress-protective enzymes. Mol Plant 3(5):927–939
12. Liscombe DK, Ziegler J, Schmidt J et al (2009) Targeted metabolite and transcript profiling for elucidating enzyme function: isolation of novel N-methyltransferases from three benzylisoquinoline alkaloid-producing species. Plant J 4:729–743
13. Vogel M, Lawson M, Sippl W et al (2010) Structure and mechanism of sanguinarine reductase, an enzyme of alkaloid detoxification. J Biol Chem 285(24):18397–18406
14. Weiss D, Baumert A, Vogel M et al (2006) Sanguinarine reductase, a key enzyme of benzophenanthridine detoxification. Plant Cell Environ 29(2):291–302
15. Donaire L, Barajas D, Martínez-García B et al (2008) Structural and genetic requirements for the biogenesis of tobacco rattle virus-derived small interfering RNAs. J Virol 82(11):5167–5177
16. Waterhouse PM, Fusaro AF (2006) Plant science. Viruses face a double defense by plant small RNAs. Science 313(5783):54–55
17. Ding SW, Voinnet O (2007) Antiviral immunity directed by small RNAs. Cell 130(3):413–426
18. Kalantidis K, Schumacher HT, Alexiadis T et al (2008) RNA silencing movement in plants. Biol Cell 100(1):13–26
19. Liu Y, Schiff M, Marathe R et al (2002) Tobacco *Rar1, EDS1* and *NPR1/NIM1* like genes are required for N-mediated resistance to tobacco mosaic virus. Plant J 30(4):415–429
20. Yellina A, Orashakova S, Lange S et al (2010) Floral homeotic C function genes repress specific B function genes in the carpel whorl of the basal eudicot California poppy (*Eschscholzia californica*). EvoDevo 1(1):1–13
21. Koncz C, Schell J (1986) The promoter of T_LDNA gene 5 controls the tissue-specific expression of chimeric genes carried by a novel type of *Agrobacterium* binary vector. Mol Gen Genet 204(3):383–396
22. Mersereau M, Pazour GJ, Das A (1990) Efficient transformation of *Agrobacterium tumefaciens* by electroporation. Gene 90(1):149–151
23. Shen WJ, Forde BG (1989) Efficient transformation of Agrobacterium spp. by high voltage electroporation. Nucleic Acids Res 17(20):8385
24. Bustin SA, Benes V, Garson JA et al (2009) The MIQE guidelines: minimum information for publication of quantitative real-time PCR experiments. Clin Chem 55(4):611–622
25. Dorak MT (2006) Real-time PCR. In: Owen E (ed) BIOS Advanced Methods. Taylor & Francis Group, New York, p 127

Chapter 8

Virus-Induced Gene Silencing Using Artificial miRNAs in *Nicotiana benthamiana*

Yang Tang, Yizhen Lai, and Yule Liu

Abstract

Virus-induced gene silencing using artificial microRNAs (MIR VIGS) is a newly developed technique for plant reverse genetic studies. Traditional virus-induced gene silencing (VIGS) assays introduce a large gene fragment, which is expressed and then converted into small RNAs by the endogenous siRNA-based gene silencing machinery of the plant host. By contrast, MIR VIGS uses well-designed miRNAs to induce RNA-mediated silencing of the target gene. Using a single artificial miRNA can provide greater specificity by reducing off-target effects. Here, we describe a detailed protocol for MIR VIGS in *Nicotiana benthamiana* using a modified *Cabbage leaf curl virus* (CaLCuV)-based vector.

Key words: Virus-induced gene silencing, MIR VIGS, Cabbage leaf curl virus, Artificial miRNA, RNA silencing

1. Introduction

Virus-induced gene silencing (VIGS) is a powerful technique to silence genes in plants (1). VIGS using artificial microRNAs (amiRNAs)(2, 3), termed "MIR VIGS," has been shown to be effective in downregulating the expression of endogenous genes (4). MIR VIGS relies on the activity of amiRNAs to induce the posttranscriptional degradation or translational repression of the target mRNA. MIR VIGS has many advantages over traditional VIGS methods. This technique is more efficient and specific in silencing both single and multiple target genes using well-designed amiRNA sequences. Unlike traditional VIGS, which generates a large variety of different 21- to 24-nucleotide siRNAs, which may cause unintended nonspecific gene suppression, amiRNAs can produce gene silencing that is more accurately targeted (2, 5).

The protocol we describe here uses a modified *Cabbage leaf curl virus* (CaLCuV)-based silencing vector. CaLCuV is a member

Annette Becker (ed.), *Virus-Induced Gene Silencing: Methods and Protocols*, Methods in Molecular Biology, vol. 975,
DOI 10.1007/978-1-62703-278-0_8, © Springer Science+Business Media New York 2013

of the genus *Begomovirus*, family *Geminiviridae*, and has a broad host range. This virus consists of two circular single-stranded DNA components, DNA-A and DNA-B, and replicates in the nucleus. DNA-A encodes coat protein, replication-associated protein, transcription activator protein, replication enhancer protein, and putative pathogenesis-related protein, and DNA-B encodes two movement proteins (6–9). The coat protein *AR1* gene is not required for viral systemic infection (8). To perform MIR VIGS using this virus, an *AR1* gene replacement DNA-A was modified to incorporate amiRNA precursor sequences, and both DNA-A and DNA-B were cloned into T-DNA cassettes for *Agrobacterium tumefaciens*-mediated delivery. Here, we describe a detailed protocol for CaLCuV-based MIR VIGS in *Nicotiana benthamiana*.

2. Materials

2.1. amiRNA Design and Virus Vector Construction

1. Target gene of interest for amiRNA design (accession number or sequence in FASTA format).
2. A computer with internet access.
3. Universal primers A (5′-CTGCAAGGCGATTAAGTTGGGTAAC-3′) and B (5′-GCGGATAACAATTTCACACAGGAAACAG-3′) and amiRNA-specific primers I, II, III, and IV, (design of these primers is described in Subheading 3.1).
4. amiRNA cloning vector pRS300 (available from Addgene, http://www.addgene.org/22846/).
5. CaLCuV cloning vectors pCVA and pCVB (to request these plasmids, please contact Yule Liu, yuleliu@mail.tsinghua.edu.cn).
6. ExTaq DNA polymerase (Takara), TIANgel Midi Purification Kit (TIANGEN), Restriction enzymes KpnI-HF (see Note 1) and XbaI (NEB), and T4 DNA ligase (Takara).
7. Equipment for PCR.

2.2. Preparation of N. benthamiana Plants

1. *N. benthamiana* seeds.
2. Nutrient soil and vermiculite.
3. Pots and clear plastic domes.
4. Light source (40 W florescent light bulb).

2.3. Agrobacterium Preparation and Infiltration

1. CaLCuV vectors pCVA, pCVB, pCVA-amiR-PDS, and pCVA-amiR-target (prepared as described in Subheading 3.1).
2. *Agrobacterium* strain GV3101.
3. LB plates containing 50 mg/L kanamycin, 30 mg/L rifampicin, and 25 mg/L gentamycin.

4. LB liquid medium containing 50 mg/L kanamycin, 30 mg/L rifampicin, and 25 mg/L gentamycin.
5. Agroinfiltration medium containing 200 μM acetosyringone (3′,5′-dimethoxy-4′-hydroxyacetophenone), 10 mM MES (2-(*N*-morpholino)ethanesulfonic acid), and 10 mM $MgCl_2$ (see Note 2).
6. *N. benthamiana* plants, approximately 4 weeks postgermination, prepared as described in Subheading 3.2.
7. 1 mL syringes with needle.
8. *Agrobacterium* culture mixture, prepared as described in Subheading 3.3.

2.4. Quantification of amiRNA and Target Gene

1. amiRNA-specific stem-loop-containing RT primer (its 5′ end is designed to target the amiRNA's last six nucleotides at the 3′ end), amiRNA-specific forward primer (containing the 5′ part sequence of the amiRNA), and universal reverse primer (5′-GTGCAGGGTCCGAGGT-3′, complementary to the stem-loop part of the RT primer).
2. Target gene mRNA specific forward primer and reverse primer.
3. Internal control (e.g., actin) -specific forward primer and reverse primer.
4. TRIzol solution (TIANGEN), RNase-free DNase I (Fermentas), M-MLV reverse transcriptase (Fermentas), and oligo-d(T).
5. Equipment for PCR.

3. Methods

In MIR VIGS, the CaLCuV *AR1* gene promoter would drive the transcription of the amiRNA precursors. The precursor sequences of amiRNAs targeting a specific gene (amiR-target) is constructed using *Arabidopsis* miR319a precursor gene as backbone and further cloned into pCVA (CaLCuV DNA-A vector) to get pCVA-amiR-target (the target is *N. benthamiana phytoene desaturase (PDS)* gene in Fig. 1). Then, the pCVA-amiR-target and pCVB (CaLCuV DNA-B vector) are transformed into *Agrobacterium* separately, and *Agrobacterium* culture mix containing pCVB and pCVA-amiR-target is injected into the plants. In the *Agrobacterium*-infected plant cell, DNA-A containing amiRNA precursor and DNA-B are produced from T-DNAs, and the inserted amiRNA precursor is transcribed in the nucleus. The amiRNA precursor is then processed to an amiRNA duplex, translocated into the cytoplasm and loaded into the RNA-induced

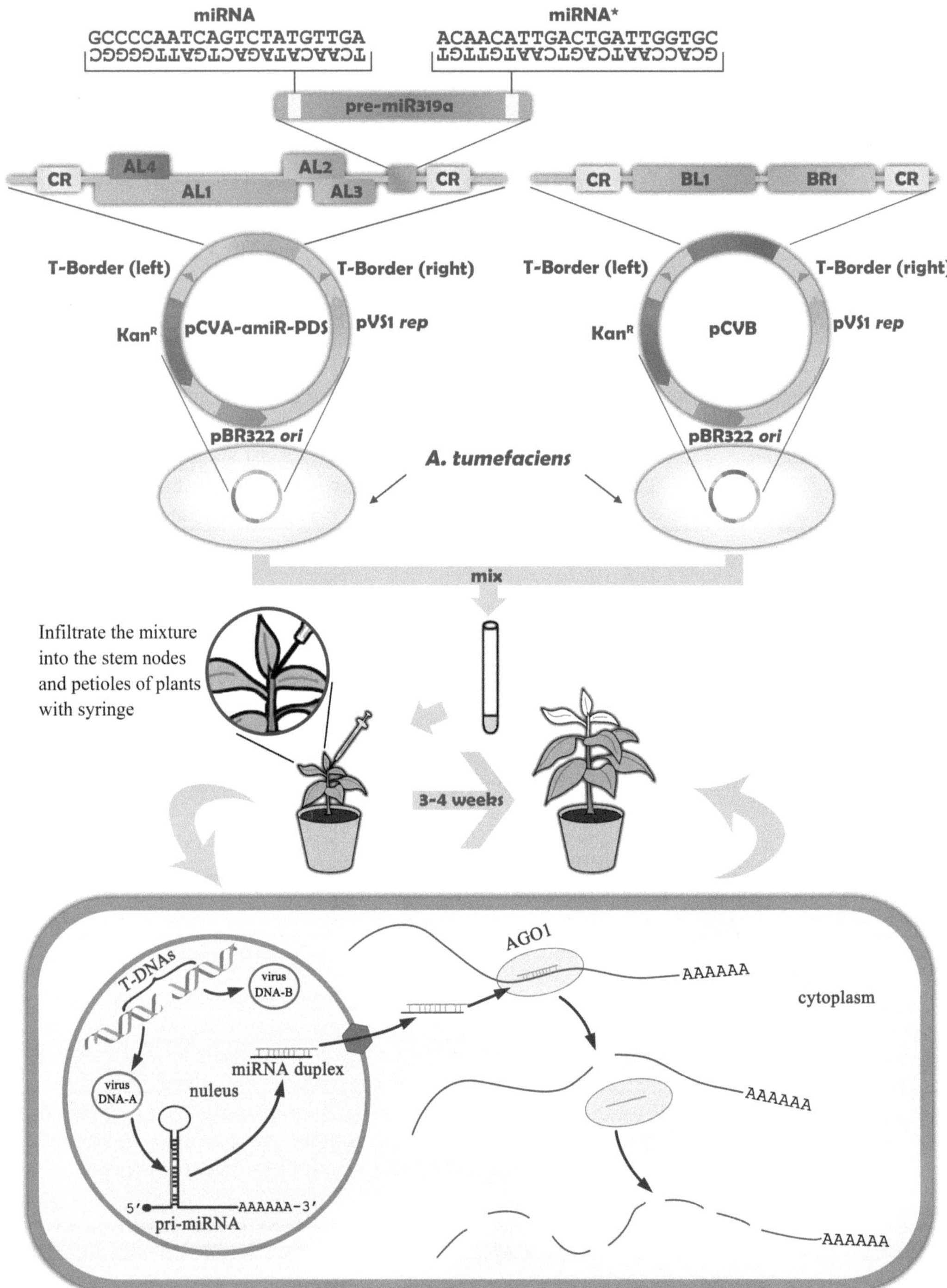

Fig. 1. Schematic of the CaLCuV-based MIR VIGS system.

silencing complex (RISC). Finally, the target gene is silenced by RISC that contains the amiRNA. Because the virus can move systemically, the target gene is silenced in all the virus-infected cells, and the corresponding gene-silencing phenotype appears throughout the plant.

3.1. amiRNA Design and Virus Vector Construction

1. An amiRNA that specifically targets the gene of interest can be designed automatically using a web application called Web MicroRNA Designer ((WMD), http://wmd3.weigelworld.org, and see Note 3, (2, 3)). On the WMD page, type the identified gene name (see Note 4) of the gene of interest (e.g., TC14913 for the *N. benthamiana PDS* gene) into the "Target genes" field. Choose the corresponding sequence library in the "Genome" field (e.g., *N. benthamiana* EST NbGI-3.0 or *Nicotiana tabacum* EST NtGI-4.0 for genes from *N. benthamiana*). If the gene is not found in the WMD transcript library (e.g., *GFP* or other nonplant genes), then input the sequence of the gene in FASTA format. Input an e-mail address in the "Email" field, and click the "Submit" button. The WMD server will process the submitted job and send an e-mail containing suggested amiRNA sequences.
2. To reduce the potential off-target effects of the amiRNA, screen the candidate amiRNA sequences using "WMD Target Search (http://wmd3.weigelworld.org/cgi-bin/webapp.cgi?page=TargetSearch;project=stdwmd)". This will find all potential target genes for the amiRNA; the candidate amiRNA that targets only the gene of interest can be chosen for further cloning.
3. Next, overlapping PCR is used to place the designed amiRNA into an endogenous miRNA precursor gene so that it can be expressed. Once you have chosen your amiRNA, use the "WMD Oligo" page (http://wmd3.weigelworld.org/cgi-bin/webapp.cgi?page=Oligo;project=stdwmd) to design the primers for amplifying the individual amiRNA gene with the backbone of the *Arabidopsis* miR319a precursor gene. Paste the chosen amiRNA sequence into the "MicroRNA sequence" field, and click the "Submit" button. Four amiRNA-specific primers (I, II, III, and IV) will be generated for each amiRNA sequence. In addition, universal primer A and primer B, which are located at 5′- or 3′-end of the miR319a precursor gene, should be prepared.
4. Use the six primers (four amiRNA special primers, I, II, III, and IV, and two universal primers, A and B) and the plasmid pRS300 to PCR amplify the amiRNA sequence (see Note 5). Two rounds of PCR are needed to produce the final amiRNA-containing construct. In the first round of PCR, three different reactions are performed simultaneously using plasmid

pRS300 as template and primer sets A and IV, II and III, or I and B, respectively. In the second round of PCR, these three PCR products are mixed (see Note 6) and used as the template to amplify the full-length amiRNA sequence using pri mers A and B. Digest the final PCR product with KpnI-HF (see Note 1) and XbaI, and ligate it into the KpnI-HF-XbaI-digested pCVA vector.

3.2. Preparation of N. benthamiana Plants

1. Distribute seeds evenly in a pot with soil, and keep the pot in a tray at 25°C under a 16 h light/8 h dark cycle. Cover the tray with a clear plastic dome until the seeds germinate (approximately 1 week).
2. Transplant seedlings individually into new pots, and grow plants under the same condition until they reach the six- to seven-leaf stage (approximately 4 weeks).

3.3. Agrobacterium Preparation and Infiltration

1. Transform the pCVA, pCVA-amiR-PDS, pCVA-amiR-target, and pCVB vectors separately into *Agrobacterium* strain GV3101 using the freeze/thaw method (10), and select for transformants on LB plates containing 50 mg/L kanamycin, 30 mg/L rifampicin, and 25 mg/L gentamycin.
2. Incubate the plates at 28°C for 2–3 days until clones appear. Pick single clones of pCVA, pCVA-amiR-PDS, pCVA-amiR-target, and pCVB vectors, and inoculate each separately into 2 mL LB liquid containing 50 mg/L kanamycin, 30 mg/L rifampicin and 25 mg/L gentamycin (because it will be used in combination with all the pCVA derivatives, the pCVB clone should be inoculated into a larger volume; use the same volume as that of all the pCVA and pCVA derivatives, that is 2× (2 + *n*) mL, where *n* is the total number of target genes).
3. Grow cultures overnight at 28°C. Pellet cells by centrifugation at 7,000 × *g* for 2 min and suspend with agroinfiltration medium to an OD_{600} of 2.0.
4. Incubate cultures at room temperature for 3–4 h. Mix each *Agrobacterium* culture containing pCVA and each pCVA derivative with culture containing pCVB, in a 1:1 ratio.
5. Load a 1 mL syringe with each *Agrobacterium* mixture, and infiltrate the mixture into the stem nodes and petioles of six to seven-leaf stage plants (see Fig. 1 and Note 7).
6. After agroinfiltration, *N. benthamiana* plants should be kept at 25°C under a 16 h light/8 h dark cycle for about 3–4 weeks until the silencing phenotype appears. For a positive control, silencing of the *phytoene desaturase* (*PDS*) gene will produce a photo-bleached phenotype on the upper leaves at approximately 3 weeks postinfiltration (Fig. 2).

Fig. 2. Phenotype produced by silencing the *PDS* gene in *Nicotiana benthamiana*. Plants were coinfiltrated with pCVA and pCVB (*left*) or pCVA-amiR-PDS and pCVB (*right*). Photographs were taken at 30 days postinfiltration.

3.4. Quantification of amiRNA and Target Gene

1. In MIR VIGS, silencing of the target gene is caused by expression of the amiRNA. There are many methods to monitor and quantify miRNAs. In our lab, we routinely use endpoint stem-loop reverse transcription (RT)-PCR (11–13) (see Note 8). Briefly, extract total RNA from leaves of silenced and nonsilenced *N. benthamiana* plants, and treat with RNase-free DNase I. Perform RT using 10 μg of total RNA, amiRNA-specific stem-loop-containing RT primers and M-MLV reverse transcriptase at 16°C for 30 min, followed by 60 cycles of pulsed RT at 30°C for 30 s, 42°C for 30 s, and 50°C for 1 s. Then, perform RT-PCR using the amiRNA-specific forward primer and the universal primer at 94°C for 2 min, followed by 30 cycles of 94°C for 15 s and 60°C for 1 min. Analyze the reaction products (about 60 bp) by electrophoresis on a 4% agarose gel in 1× Tris–acetate EDTA buffer, and quantify using appropriate imaging software.
2. The expression of the target gene can be assessed by semiquantitative RT-PCR, real-time RT-PCR, or Northern blot analysis. For semiquantitative RT-PCR, extract total RNA from leaves of silenced and nonsilenced *N. benthamiana* plants, and treat with RNase-free DNase I. Synthesize first strand cDNA from 10 μg of total RNA using oligo-d(T) primer and M-MLV reverse transcriptase. Then, perform RT-PCR using target-specific forward and reverse primers, and quantify using appropriate imaging software.

4. Notes

1. KpnI-HF is a high fidelity version of KpnI that has been engineered for reduced star activity and has high activity in the same buffer as XbaI, which simplifies the double digestion of pCVA.
2. Stock solutions are made as follows: MES and $MgCl_2$ are each dissolved separately in water to a concentration of 1 M and acetosyringone is dissolved in dimethylformamide to a concentration of 0.2 M. Dimethylformamide is a possible carcinogen, and appropriate protective measures should be adopted when using this reagent.
3. WMD is an essential, well-known tool for plant amiRNA design. A background introduction and amiRNA design guide can be found on the WMD help page, http://wmd3.weigelworld.org/cgi-bin/webapp.cgi?page=Help;project=stdwmd. In addition, some detailed protocols for amiRNA design using WMD can be found in previous volumes of "Methods in Molecular Biology" (14–17).
4. The name of the gene of interest can be found by using the BLAST tool on the WMD site. If no result is found by BLAST, then input the sequence of the gene in FASTA format into the "Target genes" field.
5. pRS300 is a pBluescript-based plasmid containing the Arabidopsis miR319a precursor gene. Any two primers flanking the MCS would be suitable as universal primers (e.g., M13F and M13R).
6. To remove background from the residual plasmid vector, we recommend gel purifying the PCR products before mixing, or using DpnI to digest residual pRS300 plasmid in the mix.
7. For each new amiRNA construct, 200–400 μL of *Agrobacterium* culture is needed for infiltration into each plant, and we usually infiltrate four plants for each construct. To avoid cross contamination, use new syringes for each construct.
8. For protocols detailing stem-loop primer design and RT-PCR, please refer to related articles in previous volumes of "Methods in Molecular Biology" (18–20).

Acknowledgments

We are very grateful to Dr. Dominique Robertson for CaLCuV vectors and to Dr. Detlef Weigel for the pRS300 vector. We also thank the National Natural Science Foundation of China (Grant nos. 31071169 and 30725002) in support of our works.

References

1. Robertson D (2004) VIGS vectors for gene silencing: many targets, many tools. Annu Rev Plant Biol 55:495–519
2. Schwab R, Ossowski S, Riester M et al (2006) Highly specific gene silencing by artificial microRNAs in Arabidopsis. Plant Cell 18:1121–1133
3. Ossowski S, Schwab R, Weigel D (2008) Gene silencing in plants using artificial microRNAs and other small RNAs. Plant J 53:674–690
4. Tang Y, Wang F, Zhao J et al (2010) Virus-based microRNA expression for gene functional analysis in plants. Plant Physiol 153: 632–641
5. Schwab R, Palatnik JF, Riester M et al (2005) Specific effects of microRNAs on the plant transcriptome. Dev Cell 8:517–527
6. Hill JE, Strandberg JO, Hiebert E et al (1998) Asymmetric infectivity of pseudorecombinants of cabbage leaf curl virus and squash leaf curl virus: implications for bipartite geminivirus evolution and movement. Virology 250: 283–292
7. Paximadis M, Idris AM, Torres-Jerez I et al (1999) Characterization of tobacco geminiviruses in the Old and New World. Arch Virol 144:703–717
8. Turnage MA, Muangsan N, Peele CG et al (2002) Geminivirus-based vectors for gene silencing in Arabidopsis. Plant J 30:107–114
9. Muangsan N, Robertson D (2004) Geminivirus vectors for transient gene silencing in plants. Methods Mol Biol 265:101–115
10. Wise AA, Liu Z, Binns AN (2006) Three methods for the introduction of foreign DNA into Agrobacterium. Methods Mol Biol 343: 43–53
11. Chen C, Ridzon DA, Broomer AJ et al (2005) Real-time quantification of microRNAs by stem-loop RT-PCR. Nucleic Acids Res 33:e179
12. Tang F, Hajkova P, Barton SC et al (2006) MicroRNA expression profiling of single whole embryonic stem cells. Nucleic Acids Res 34:e9
13. Varkonyi-Gasic E, Wu R, Wood M et al (2007) Protocol: a highly sensitive RT-PCR method for detection and quantification of microRNAs. Plant Methods 3:12
14. Khraiwesh B, Fattash I, Arif MA et al (2011) Gene function analysis by artificial microRNAs in Physcomitrella patens. Methods Mol Biol 744:57–79
15. Manavella PA, Rubio-Somoza I (2011) Engineering elements for gene silencing: the artificial microRNAs technology. Methods Mol Biol 732:121–130
16. Schwab R, Ossowski S, Warthmann N et al (2010) Directed gene silencing with artificial microRNAs. Methods Mol Biol 592:71–88
17. Eamens AL, Waterhouse PM (2011) Vectors and methods for hairpin RNA and artificial microRNA-mediated gene silencing in plants. Methods Mol Biol 701:179–197
18. Varkonyi-Gasic E, Hellens RP (2010) qRT-PCR of Small RNAs. Methods Mol Biol 631:109–122
19. Chen C, Tan R, Wong L et al (2011) Quantitation of microRNAs by real-time RT-qPCR. Methods Mol Biol 687:113–134
20. Varkonyi-Gasic E, Hellens RP (2011) Quantitative stem-loop RT-PCR for detection of microRNAs. Methods Mol Biol 744: 145–157

Chapter 9

The Use of VIGS Technology to Study Plant–Herbivore Interactions

Ivan Galis, Meredith C. Schuman, Klaus Gase, Christian Hettenhausen, Markus Hartl, Son T. Dinh, Jianqiang Wu, Gustavo Bonaventure, and Ian T. Baldwin

Abstract

Plants employ a large variety of defense strategies to resist herbivores, which require transcriptional reprogramming of cells and profound changes in plant metabolism. Due to the large number of genes involved in defense processes, rapid screening strategies are essential for elucidating the contributions of individual genes in the responses of plants to herbivory. However, databases and seed banks of mutant plants which allow rapid retrieval of mutant genotypes are limited to a few model plant species, namely, *Arabidopsis thaliana* and *Oryza sativa* (rice). In other plants, virus-induced gene silencing (VIGS) offers an efficient alternative for screening the functions of individual genes in order to prioritize the allocations of the large time investments required to establish stably transformed RNAi-silenced lines. With VIGS, it is usually possible to achieve strong, specific silencing of target genes in the ecological models *Nicotiana attenuata* and *Solanum nigrum*, allowing the rapid assessment of gene silencing effects on phytohormone accumulation, signal transduction and accumulation of defense metabolites. VIGS plants are also useful in bioassays with specialist and generalist herbivores, allowing direct verification of gene function in plant resistance to herbivores.

Key words: *Nicotiana attenuata*, *Solanum nigrum*, Post-transcriptional gene silencing (PTGS), Herbivory, Plant hormone level measurements, Secondary metabolite profiling

1. Introduction

Originally designed for the wild tobacco species *Nicotiana benthamiana* (1), VIGS has emerged as an extremely powerful functional genomics tool for knocking down the expression of target genes in plants. We adopted and gradually optimized the *N. benthamiana* VIGS protocol to perform rapid screens of gene

Annette Becker (ed.), *Virus-Induced Gene Silencing: Methods and Protocols*, Methods in Molecular Biology, vol. 975, DOI 10.1007/978-1-62703-278-0_9, © Springer Science+Business Media New York 2013

function in a diploid species of tobacco, *Nicotiana attenuata*, which has been developed over the past 20 years as an ecological model plant. *N. attenuata* is native to the Great Basin Desert of the southwestern USA. In its natural environment, *N. attenuata* is attacked by a variety of herbivores which have exerted strong selection for the evolution of efficient defense mechanisms against insect attackers. Post-transcriptional gene silencing (PTGS) used as a tool to dissect the roles of particular genes in the plant's adaptation to herbivory has revealed the ecological function of, and mechanisms behind multiple defense systems in *N. attenuata* (2–12).

Many of these studies have used VIGS for initial screening of candidate gene function to be followed by the elucidation of phenotypes in stably silenced lines created via integration into the genome, and expression of antisense or inverted repeat fragments of genes of interest *in planta*. Typically, VIGS experiments have provided the first information about the function of these genes, and this information has provided the foundation for novel hypotheses and accelerating research on herbivory-related genes in *N. attenuata*. The use of VIGS, for example, significantly contributed to elucidating the roles of MAP kinase signaling in defense against herbivores (7), the defensive role of 17-hydroxygeranyllinalool diterpenoid glycosides (HGL-DTGs) (8) and the role of phenylpropanoid-polyamine conjugates (PPCs) (12) in the plant's defense against chewing herbivores. In addition, we extended this technology to another solanaceous plant model, *Solanum nigrum*, to successfully knock down gene expression; VIGS helped to reveal the role of protease inhibitors (PIs) (13) and leucine aminopeptidase (14) in the defense of black nightshade plants against insect herbivores.

The VIGS vectors used for *N. attenuata* and *S. nigrum* are based on the pBINTRA6/pTV00 tobacco rattle virus (TRV) system developed by Ratcliff and colleagues (1), and the enhanced TRV system based on pTRV1/pTRV2 vectors developed by Liu and colleagues (15), respectively (see Note 1). TRV is a type member of the tobravirus group that has a bipartite single-stranded positive-sense RNA genome, consisting of RNA-1 and RNA-2. The two genome parts are contained in rod-shaped, predominantly long (RNA-1) or short (RNA-2) particles (16). The replication of the viral RNA involves a double-stranded RNA (dsRNA), synthesized by an associated RNA-dependent RNA polymerase (RdR; replicase), as an intermediate in the cycle between parental and progeny genomes (17). TRV has several distinct advantages over other viruses developed for VIGS (potato virus X, PVX; tobacco mosaic virus, TMV; tobacco golden mosaic virus, TMGV) including relatively mild symptoms of infection, infection of large patches of neighboring cells, and migration to growing meristems and thus efficiently into new tissues in all parts of the plant (1).

To enable VIGS in plants, the cDNA of both parts of the TRV genome has been modified and cloned separately on the T-DNA of two binary plant transformation vectors: RNA-1 on pBINTRA6 (20.2 kb) and RNA-2 on pTV00 (5.6 kb) (see Note 2). pBINTRA6

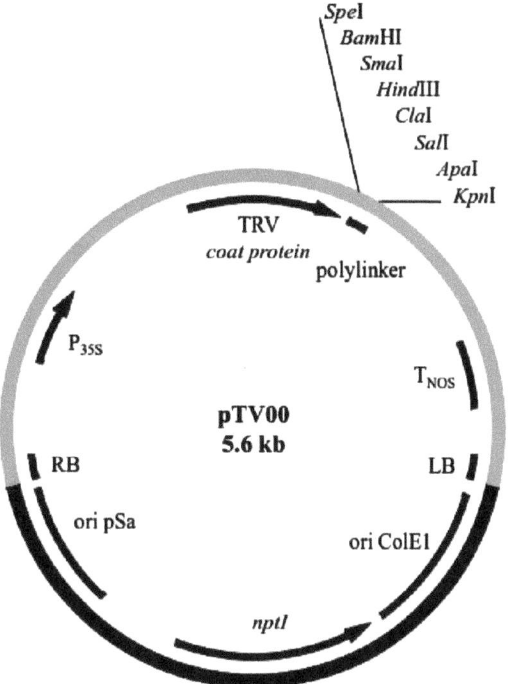

Fig. 1. Binary plant transformation vector pTV00. *LB/RB*: left/right border of T-DNA, P_{35S}: 35S promoter from cauliflower mosaic virus, T_{NOS}: terminator of the nopaline synthase gene from the Ti plasmid of *Agrobacterium tumefaciens*, *npt*I: aminoglycoside phosphotransferase class I, *ori*: origin of replication.

contains a full length infectious TRV cDNA consisting of the coding sequences for TRV RdR, movement protein, and the 16 kDa protein. The RdR open reading frame was interrupted by an intron from *Arabidopsis* to prevent the expression of a protein that is toxic to *Escherichia coli*. The complete RNA-1 is driven by a cauliflower mosaic virus (CaMV) 35S promoter and terminated by a CaMV 35S terminator. The TRV RNA-2 on pTV00 (Fig. 1) only consists of the coat protein gene; the nonessential 29.4 kDa and 32.8 kDa protein genes have been replaced by a 50 bp polylinker (*Spe*I to *Kpn*I; Fig. 1). RNA-2 and the polylinker sequence are co-transcribed under the control of a CaMV 35S promoter and a NOS terminator.

To silence a gene of interest (*goi*), a 150–400 bp antisense fragment of this gene is cloned in the polylinker of pTV00 (see Note 3). In our experience, it is possible to silence two independent genes by tandem insertion of two 150–200 bp sequences specific for each *goi*. RNAi stably silenced lines of one gene also can be used as starting material for a successful VIGS of another independent target *goi* (13). After inoculation of *N. attenuata* with a 1:1 mixture of both *Agrobacterium* cultures, carrying, respectively, pTV and pBINTRA6 vectors, the T-DNA of both binary plant transformation vectors is integrated into the genome of the infected tissue, which results in the production of both viral RNA species and, later, in virus replication. Since RNA silencing is ubiquitously triggered by dsRNA (18), and the fragment of the *goi* is co-transcribed with RNA-2, VIGS of this gene will be induced independently of the orientation of the cloned fragment; however, antisense fragments were shown to result in slightly more efficient gene silencing in *N. attenuata* (19). Importantly, VIGS spreads to both root and shoot tissues which have developed after inoculation (19).

Positive and negative VIGS controls are necessary, and easily integrated into each experiment. The empty cloning vector pTV00 can be used as a negative control to exclude nonspecific phenotypic effects of VIGS (see Note 4). The positive control vector pTVPD was constructed by cloning a 206 bp fragment of the *N. benthamiana phytoene desaturase* gene (pds) in antisense orientation in the polylinker of pTV00 (provided by the lab of Sir David Baulcombe). Phytoene desaturase oxidizes and cyclizes phytoene to α- and β-carotene, important intermediates of the antenna pigments of the photosystems of plants. Silencing of *pds* results in visible bleaching of green plant tissues and is therefore an ideal indicator to assess the success of VIGS (1, 19).

In this chapter, we provide guidelines for inducing VIGS in *N. attenuata* and *S. nigrum* plants as a basic tool for screening the function of herbivory-associated plant genes. Such a candidate gene may be, for example, a novel transcription factor obtained from a microarray experiment comparing herbivore-attacked and control leaf tissues. As the mode of action of such a gene is usually unknown, we designed a rapid VIGS protocol complemented with a broadly targeted screening procedure that includes determination of silencing efficiency, measurement of the levels of plant hormones such as jasmonates, salicylic and abscisic acids, profiling of secondary metabolites, and herbivory bioassays that can be all conveniently incorporated into a single VIGS experiment.

2. Materials

2.1. Materials for Preparation of Bacterial Clones

2.1.1. Cloning of Gene Fragments into VIGS Vectors

1. Restriction enzymes and buffers; DNA ligase kit; PCR polymerase kit; PCR primers.
2. 5 M NaCl.
3. 70% and 96% ethanol.
4. Nuclease-free water.
5. *E. coli* competent cells.
6. Sterile liquid LB medium and LB agar with antibiotics (50 mg/L kanamycin).

7. Gel and plasmid DNA extraction kits (e.g., MN NucleoSpin Extract II; MN NucleoSpin Plasmid; http://www.mn-net.com).
8. Agarose gel electrophoresis.
9. PCR, microcentrifuge, and electroporation instruments.
10. Shaker for cultivation of bacterial cultures at 37°C.

2.1.2. Preparation of *A. tumefaciens* GV3101 Electrocompetent Cells

1. Sterile YEP medium (10 g/L yeast extract, 10 g/L peptone, 5 g/L sodium chloride; adjusted to pH 7) (2 × 0.52 L) with antibiotics.
2. Sterile 10% (v/v) glycerol (4 × 0.5 L).
3. Sterile conical flasks/Erlenmeyer flasks (2 × 1.0 L, 1 × 0.1 L).
4. Sterile Oak Ridge centrifuge tubes, PC, Nalgene 31180085 85 mL (6×).
5. Sterile 1.5 mL tubes.
6. Spectrophotometer for determination of bacterial density.
7. Refrigerated centrifuge.
8. Shaker for cultivation of bacterial cultures at 28°C.
9. Freezer (−80°C).

2.1.3. Transformation of *A. tumefaciens* GV3101

1. Sterile electroporation cuvette, 1 mm gap.
2. 1 mL sterile S.O.C. (e.g., from http://www.invitrogen.com).
3. 0.3 mL sterile 80% glycerol.
4. Sterile liquid LB medium and LB agar with antibiotics.
5. Electroporation instrument.
6. Shaker for cultivation of bacterial cultures at 28°C.
7. Microcentrifuge.

2.2. Materials for Induction of VIGS in *N. attenuata*

2.2.1. Materials for Germinating *N. attenuata* Seeds

1. *N. attenuata* Torr. ex S. Watson (31st inbred line) seeds, 40–50 seeds per construct of interest (see Note 5).
2. Seed sterilization solution: 2% (w/v) of dichloroisocyanuric acid sodium salt with 0.005% Tween-20 in dH_2O (prepare a stock of 0.5% Tween-20 to dilute).
3. Stock solution of 0.1 M GA_3 (gibberellic acid) in ethanol, filter sterilized.
4. 50× diluted liquid smoke (House of Herbs, Passaic, NJ), in dH_2O, autoclaved.
5. Sterile dH_2O.

6. Petri dishes with 0.6% plant agar- or phytagel-supported Gamborg's B5 cultivation media (1× strength).
7. Growth chamber for plates with seedlings.

2.2.2. Materials for Bacteria Cultivation and Plant Inoculation

1. Constructs: pBINTRA6 (RNA1), pTV (RNA2 ±*goi*), pTVPD (PDS fragment from *N. benthamiana*) in *A. tumefaciens* strain GV3101. Plates for *A. tumefaciens* pre-cultures (LB or YEP with 50 mg/L kanamycin).
2. *N. attenuata* plants in young rosette stage, growing in a climate chamber at a constant temperature program of 20°C, 16 h day/8 h night, ca.65% relative humidity and low light intensity (115–165 μmol s^{-1} m^{-2}PAR).
3. YEP medium: 10 g/L yeast extract, 10 g/L peptone (from soy), and 5 g/L sodium chloride, autoclaved; LB medium can also be used as a substitute for YEP, but *A. tumefaciens* grows more slowly in LB.
4. 1,000× kanamycin stock solution in dH_2O, filter-sterilized (50 mg/mL).
5. 10× stock solutions of 100 mM $MgCl_2$ and 100mM2-(*N*-morpholino) ethanesulfonic acid (MES) (adjust pH of MES to 5.5–6.0 with KOH), autoclaved.
6. Erlenmeyer flasks for growing *Agrobacterium* cultures, autoclaved.
7. Shaker for cultivation of bacterial cultures in flasks at 28°C.
8. Spectrophotometer for determination of bacterial density.
9. 1 mL syringes without needle.
10. Centrifuge.

2.3. Materials for Induction of VIGS in S. nigrum

2.3.1. Materials for Growing Sterile S. nigrum Seedlings

1. *S. nigrum* seeds (see Note 6), 30–40 seeds per construct of interest.
2. Seed sterilization solution: 2% (w/v) of dichloroisocyanuric acid sodium salt with 0.005% Tween-20 in dH_2O (prepare a stock of 0.5% Tween-20 to dilute).
3. Seed stratification medium: sterile 1 M KNO_3.
4. Cultivation medium: sterile standard Gamborg's B5 medium in petri dishes (0.6% plant agar).
5. Sterile dH_2O.
6. Growth chamber for plates with seedlings.

2.3.2. Materials for Bacteria Cultivation and Plant Inoculation

1. Constructs: pTRV1 (RNA1), pTRV2 (RNA2 +*goi*), pTRV2-SnPDS (PDS fragment from *S. nigrum*) in *A. tumefaciens* strain GV3101 (14, 15).

2. *S. nigrum* wild-type plants in seedling stage (7–10 days after sowing).
3. 9×9×9.5 cm pots filled with moist peat-based substrate.
4. 1,000× kanamycin stock solution in dH_2O, filter-sterilized (50 mg/mL).
5. 20 mM acetosyringone in 50% ethanol, filter-sterilized.
6. LB agar plates, 50 mg/L.
7. LB liquid medium with 1.95 g 2-(*N*-morpholino) ethanesulfonic acid (MES)/L, 50 mg kanamycin/L, 20 μM acetosyringone: prepare 400 mL per construct of interest (200 mL for construct of interest and 200 mL for pTRV1), plus an extra amount for pre-cultures and dilutions (depending on number of constructs, minimum 200 mL). A smaller amount of pTRV2-SnPDS culture for inoculation of 5–10 plants should also be prepared. Add sterile filtered kanamycin and acetosyringone after autoclaving only.
8. 10× infiltration medium: 100 mM $MgCl_2$, 100 mM MES (adjust pH to between 5.5 and 6.0 with KOH). For use, dilute 1:10 with dH_2O, add 10 mL of the dilution per L of acetosyringone stock (200 μM final concentration).
9. Sterile Erlenmeyer flasks (500 mL), one for each construct of interest and one for each accompanying pTRV1 culture. Add one smaller 100 mL bottle for the pTRV2-SnPDS culture.
10. 100 mL glass beakers, one per construct of interest (including the PDS positive control).
11. Desiccator and vacuum pump (capable of reaching 60–100 mbar).
12. Spectrophotometer for determination of bacterial density.
13. Refrigerated centrifuge.

2.4. Materials for Functional Screening of VIGS Plants

2.4.1. Plant Elicitations

Preparation of elicitors is described in Subheading 3.4 Plant elicitations (a).

1. Lepidopteran larvae (*Manduca sexta*, *Spodoptera littoralis* or *exigua*) for collecting herbivore oral secretions (OS).
2. dH_2O.
3. Tracing (pattern) wheel with serrated edge.
4. 20 μL pipette and tips.
5. Support for leaf to be treated (e.g., plastic disc, ca. 5–10 cm diameter).
6. Methyl jasmonate (MeJA) in pesticide-free lanolin, 150 μg per 20 μL.

2.4.2. Quantitative RT-PCR

1. Liquid nitrogen for freezing and grinding of samples.
2. Plant total RNA isolation kit (or TRIzol Reagent [Invitrogen] and solutions for purification of TRIzol-isolated total RNA).
3. RNAse-free water (e.g., DEPC-treated water).
4. cDNA synthesis kit supplied with oligo(dT) or random hexamer primers.
5. Standard cDNA for calibration curve (see Note 7).
6. qRT-PCR instrument, PCR tubes or plates with optically transparent lids for qRT-PCR.
7. Spectrophotometer for determination of RNA concentrations.
8. Microcentrifuge.

2.4.3. Analysis of Phytohormones

1. Liquid nitrogen for freezing and grinding of samples.
2. Mortars and pestles or mechanical grinder (ball mill).
3. Extraction solvent: ethyl acetate (HPLC grade).
4. 70/30 (v/v) methanol–deionized water (HPLC grade).
5. Isotopically labeled phytohormone standards (IS; see Table 2).
6. LC-MS/MS instrument equipped with an analytical column.
7. LC-MS/MS solvents: 0.05% formic acid in water (solvent A); methanol HPLC grade (solvent B).
8. Vacuum rotary evaporator (Speed Vac) with trap for organic solvents.
9. Refrigerated microcentrifuge.

2.4.4. Analysis of Secondary Metabolites

1. Liquid nitrogen for freezing and grinding of samples.
2. Mortars and pestles or mechanical grinder (ball mill).
3. Extraction solvent: 40/60 (v/v) methanol–dH_2O; acidified with 0.1% acetic acid.
4. Nicotine, chlorogenic acid, and rutin calibration standards (10–250 ng/μL range) dissolved in 40/60 (v/v) methanol–dH_2O.
5. HPLC instrument with an analytical reverse phase (RP) column and UV detector (see Note 8).
6. HPLC solvents: HPLC-grade methanol (solvent A); 0.1% acetic acid; 0.1% ammonia (concentrated) in dH_2O (solvent B).
7. Microcentrifuge.

2.4.5. Herbivore Bioassays

1. VIGS *N. attenuata* plants using *goi* and empty vector (EV) constructs.
2. *M. sexta* eggs or neonates (freshly hatched larvae).

3. *S. littoralis* neonates reared on artificial diet for several days (5–10 mg caterpillars).
4. Perforated plastic bags, soft wire or string to close the open ends.
5. Soft metal forceps.
6. Analytical balances (accuracy 0.1 mg).

3. Methods

3.1. Construction of VIGS Vectors (see Note 9)

3.1.1. Cloning of Gene Fragments in VIGS Vector

A fragment of the *gene of interest* (*goi*) to be silenced must be cloned in the polylinker of pTV00 (Fig. 1); routinely, we do this in antisense orientation after choosing appropriate unique cloning sites. Because of enzyme buffer compatibility, we routinely use the *Bam*HI and *Sal*I sites for cloning, but another combination of restriction sites can be used if *Bam*HI or *Sal*I sites cannot be avoided on the *goi* fragment.

1. Find a region of sufficient size in the *goi* cDNA sequence that does not contain the chosen restriction sites. Because of better representation in cDNA samples, the 3′ end is favored. If only one member of a gene family should be silenced, the part having the least identity with other family members should be chosen. For silencing of homologous genes, choose the region with the highest identity (see Note 3).
2. Design the PCR primer pair for the amplification of a *goi* PCR fragment from the chosen region at an appropriate size (150–400 bp) (see Note 3). The primers should carry a 5′ GC rich sequence followed by the restriction site of one of the cloning enzymes and sufficient nucleotides for annealing to the *goi* (about 21–25 nucleotides, same GC content for both primers and a GC clamp at the 3′ end, no primer dimers possible). Here is an example of a primer pair for *Bam*HI–*Sal*I-cloning: forward primer (*Sal*I site underlined): 5′-GCGGCG<u>GTCGAC</u>($Ngoi_{20\text{-}24}$)G-3′; reverse primer (*Bam*HI site underlined): 5′-GCGGCG<u>GGATCC</u>($Ngoi_{20\text{-}24}$)G-3′.
3. PCR amplify the desired *goi* fragment with the primers using cDNA prepared from plant tissues expressing the *goi*. Preferably, use a proofreading polymerase and follow the manufacturer's instructions for optimal performance. Pipette a 50 μL reaction and divide it into 10 μL aliquots. Run each aliquot at a different annealing temperature in the range of 55–65°C.
4. Run your PCR products along with a molecular mass standard on a 1% agarose gel. Excise PCR fragments of the expected size and extract the DNA with an appropriate kit (e.g., MN NucleoSpin Extract II, http://www.mn-net.com). About 200 ng of PCR fragment will be needed for further cloning. If the amount of the

PCR fragment is too low, repeat the PCR at the optimal annealing temperature. If the PCR result is unsatisfying at all annealing temperatures, design a new primer pair.

5. Digest, in separate reactions, 200 ng of pTV00 vector and 200 ng of the PCR fragment with the restriction enzymes according to the manufacturer's instructions.
6. Run both digestions using a 1% agarose gel; excise the 5.5 kb pTV00 band and the digested PCR fragment. Extract DNA.
7. Set up a 20 μL ligation reaction with approximately 50 ng of digested pTV00 and 100 ng digested PCR fragment. Ligate overnight at 16°C.
8. Add 2 μL 5 M NaCl and 50 μL ethanol, mix, and precipitate for at least 2 h in a deep freezer (−80°C).
9. Pellet the precipitated DNA by 10 min centrifugation at 16,000 × *g*. Wash the pellet twice with 70% ethanol, and spin down the pellet for 1 min at the same speed. Carefully remove the supernatant, briefly dry the pellet, and dissolve the pellet in 20 μL of nuclease-free water.
10. Use electroporation to transform an *E. coli* strain carrying lacIq (e.g., TOP10F′, http://www.invitrogen.com) with the purified ligation mixture (see Note 10).
11. Plate the transformed cells on LB plates containing 50 mg/L kanamycin and incubate the bacteria at 37°C overnight. Expect 1/100 of the number of colonies you would obtain with *E. coli* standard cloning vectors (like pUC19 with ampicillin selection).
12. Use few single colonies to inoculate per colony 5 mL of LB liquid medium containing 50 mg/L kanamycin. It works well to use a sterile forceps to hold a sterile plastic pipette tip, touch the side of the pipette tip to colonies on the plate, and drop the tip into the culture tube with your LB. Shake overnight at 200 rpm at 37°C.
13. Isolate plasmid DNA using a standard kit (e.g., MN NucleoSpin Plasmid).
14. Digest 200 ng of each plasmid with the cloning enzymes and run the fragments on a 1% agarose gel. Plasmids that contain both the correct pTV00 and *goi* fragment are used to confirm the correct sequence of the insert by sequencing with the primers:

    TRV FOR 5′-GCTGCTAGTTCATCTGCAC-3′.
    TRV REV 5′-GCACGGATCTACTTAAAGAAC-3′.
15. One plasmid with the correct sequence is used to transform *A. tumefaciens* electrocompetent cells in the following section.

3.2. Transformation of A. tumefaciens Cells

3.2.1. Preparation of A. tumefaciens GV3101 Electrocompetent Cells

1. Inoculate 20 mL YEP-medium containing 50 mg/L rifampicin and 7.5 mg/L tetracycline with *A. tumefaciens* GV3101 and incubate on a shaker overnight at 200 rpm and 28°C.
2. Inoculate 2 × 0.5 L YEP medium in sterile 1 L conical flasks each with 10 mL of the pre-culture. Let the bacteria grow at 200 rpm and 28°C for about 8 h until they reach an A_{600nm} of 0.5–0.7 (logarithmic growth phase). During aliquoting and monitoring of growth, remove cells from shaker as briefly as possible.
3. Fill the bacterial cultures into 6 × 85 mL Oak Ridge tubes (maximum 56 mL volume) and spin them in a fixed angle rotor at 2,400 × *g* for 5 min at 4°C.
4. Pour off the supernatant, thereby preserving the pellet. Refill the tubes with the same volume of the remaining bacterial culture. Spin the bacterial cultures in a fixed angle rotor at 2,400 × *g* for 5 min at 4°C. Keep cell suspensions or pellets on ice between each centrifugation step!
5. Repeat step 5 to pellet all bacterial culture.
6. Carefully remove all supernatant. If necessary, briefly spin the tubes again after emptying and carefully remove the last drops with a pipette tip. It is important to remove as much salt as possible to obtain cells with lowest possible conductivity for highest possible competence.
7. Resuspend the cells in 56 mL pre-cooled 10% glycerol per tube (starting with 10 mL, resuspend the pellet gently without vortexing and then add the remaining 10% glycerol).
8. Spin for 10 min and remove the supernatant as before. Continue with at least three cycles of washing as follows.
9. Resuspend the cells in 56 mL 10% glycerol. Centrifuge and remove supernatant as before. Resuspend each pellet in 18 mL 10% glycerol. Combine the bacterial suspensions from three tubes into one tube.
10. Resuspend each pellet in 28 mL 10% glycerol. Combine the bacterial suspensions from the two remaining tubes into one tube. Centrifuge as above. Remove the supernatant as carefully (and as completely) as possible.
11. Resuspend the final pellet in 1.5 mL 10% glycerol and aliquot 40 μL per 1.5 mL microcentrifuge tube.
12. Freeze the aliquots at −80°C.

3.2.2. Transformation of A. tumefaciens GV3101 by Electroporation

1. Pre-warm 1 mL of S.O.C. per construct to 28°C.
2. Slowly thaw on ice one aliquot (40 μL) of competent *A. tumefaciens* GV3101 cells.
3. After thawing, mix the cells with 1 μL (about 100–200 ng) of plasmid DNA from a correct pTV*goi* clone.

4. Fill the mixture into an ice-cold electroporation cuvette with a gap of 1 mm.
5. Perform electroporation using an electroporator (e.g., Bio-Rad MicroPulser, http://www.bio-rad.com) at 2.5 kV for 5 ms, or use the instrument preset for *E. coli*.
6. Add 1 mL of S.O.C. pre-warmed to 28°C immediately after pulse.
7. Mix the cells with the S.O.C., pipette the suspension into a 1.5 mL microcentrifuge tube, and incubate for 30 min at 28°C.
8. Spin down the cells for 1 min at 3,000 × *g*.
9. Carefully remove supernatant but leave about 100 μL of medium in the tube.
10. Resuspend the cells in the remaining medium and plate 10 and 90 μL, respectively, on two LB agar plates containing 50 mg/L kanamycin.
11. Incubate the plates for 48 h at 26°C. Expect ca. 100 and 1,000 colonies on the plates depending on the volume of transformation mix used.
12. Isolate a single colony. Use this colony to inoculate 3 mL of liquid LB medium containing 50 mg/L kanamycin. Grow the bacteria in a shaker at 200 rpm and 28°C for about 20 h until they reach an A_{600nm} of about 1.
13. Prepare a glycerol stock by mixing 0.7 mL of this culture with 0.3 mL of 80% glycerol and freeze at −80°C. The culture or the glycerol stock can be used as starting culture for the VIGS procedure.

3.3. VIGS Inoculation Methods

3.3.1. *N. attenuata* Inoculation Protocol

(a) Germination of *N. attenuata* seeds:

1. *N. attenuata* Torr. Ex S. Watson (31st inbred line) seeds, originally collected from a native population at a field site located in Utah (USA), are used in the experiment (see Note 5).
2. Prepare fresh sterilization solution containing 2% (w/v) of dichloroisocyanuric acid sodium salt and 0.005% Tween-20 in dH_2O.
3. Incubate the seeds in 5 mL of sterilization solution for 5 min using 15 mL conical tube; shake occasionally. The following steps should be carried out under a sterile hood.
4. Decant solution and wash seeds in 5–10 mL sterile dH_2O. Repeat decanting and washing steps at least three times.
5. Add 50 μL of 0.1 M GA_3 stock to 5 mL of diluted smoke solution and incubate seeds for 1 h in this solution (see

Note 11). Shaking occasionally or inclining the tubes to provide a larger surface area of seed exposure during incubation yields more even germination and greater germination efficiency.

6. Decant smoke/GA_3 solution, wash seeds three times with sterile dH_2O, and distribute 30–40 seeds/plate with 0.6% agar or phytagel-supported Gamborg's B5 cultivation media.
7. Allow plates to dry briefly until any water transferred with seeds has evaporated, then close with Parafilm (Pechiney Plastic Packaging Company, Chicago, IL) and transfer to a growth chamber (16 h light/8 h dark, 26°C, 155 μmol $s^{-1}m^{-1}$ PAR at shelf height).
8. After 10 days carefully remove seedlings from agar and transfer to soil in small Teku plastic pots with Klasmann plug soil (Klasmann-Deilmann GmbH, Geesten, Germany).
9. After 10 days in Teku pots, transfer seedlings to 1 L pots in soil (0.75 g Superphosphate, Multimix 14:16–18 [Haifa Chemicals Ltd., Haifa Bay, Israel], 0.35 g $MgSO_4 \cdot 7H_2O$ [Merck KGaA, Darmstadt, Germany], 0.05 g Micromax [Scotts Deutschland GmbH, Nordhorn, Germany] per 1 L). Place pots in groups of 10–11 into flat plastic trays with a ca. 5.5 cm rim; this is convenient for the VIGS inoculation. Transfer to a climate chamber with a constant temperature of 20°C and 16 h day/8 h night light regime, ca. 65% relative humidity, and low light intensity (115–165 μmol s^{-1} m^{-2}PAR).

(b) Inoculation of *N. attenuata* seedlings by syringe infiltration:

1. Plants should be inoculated at 3–5 days post-potting = 23–25 days post-germination (see Note 12). It is very important that plants are in the right stage because with older plants, VIGS has been found to be less efficient.
2. The whole experiment is carried out in a climate chamber with a constant temperature of 20°C and 16 h day/8 h night light regime, ca.65% relative humidity, and low light intensity (115–165 μmol s^{-1} m^{-2}PAR). For timing and preparation of cultures, follow the schematic diagram shown in Fig. 2. Remember that proper timing is essential for successful establishment of VIGS.
3. Ca. 3 days before plants are potted, streak *A. tumefaciens* cultures onto a YEP or LB agar plate with 50 mg/L Kan. Allow to grow for 2 days at 28°C. You can store the plates at 4°C for approximately 1 month.

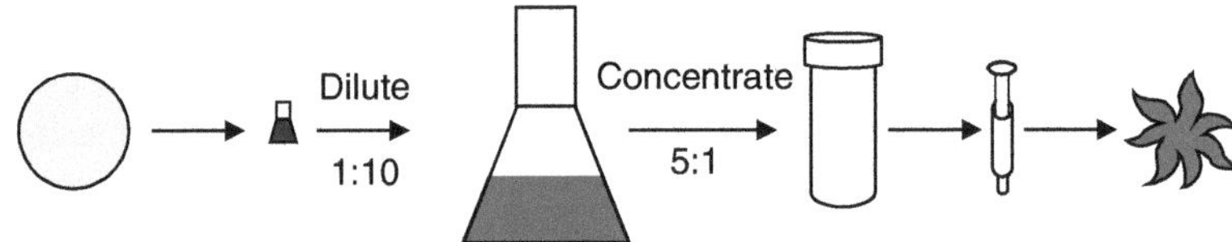

What	Bacterial plate	Pre-culture	YEP liquid culture	Inoculation solution MES/$MgCl_2$	1 mL syringe	Plant
When	3 days before plants potted	1 day before VIGS	Early morning of VIGS or evening before VIGS	Day of VIGS	Day of VIGS	Germinate 23-25 days before VIGS
How many/ much	1/culture	1/culture	≥1/culture	≥1 mL/ plant	≥1 pcs/ culture	As required

Fig. 2. Schematic overview of preparation of *A. tumefaciens* cultures required for the *N. attenuata* inoculation protocol. The timing of cultures needs to be carefully coordinated with the growth of *N. attenuata* plants in growth chamber as the use of too young (or too old) plants will result in poor VIGS efficiency.

4. On or before the day plants are potted, prepare Erlenmeyer flasks and enough YEP media (without Kan) for each construct and autoclave. For every 1 mL of inoculation solution, you will need 2.5 mL of liquid YEP culture with pTV*goi* plus 2.5 mL pBINTRA liquid YEP culture; calculate 1 mL inoculation solution per plant (empirical formulas to calculate proper amounts of cultures necessary for a typical experiment are provided in Table 1; do not forget to prepare a small amount of pTVPD culture for pds silencing as your positive control). Also prepare small flasks for the pre-culture if necessary. Do not fill flasks to more than 50% of their maximum volume to allow sufficient gas exchange while shaking cultures.
5. On day 2 after potting, check the size of the plants. If the plants are too small, postpone the VIGS experiment for 1–2 days; ideally, the leaf size should be 2–3 cm in length during infiltration. If the size is sufficient, start small overnight pre-cultures of *A. tumefaciens* in small Erlenmeyer flasks or 15 mL culture tubes with YEP + Kan 50 mg/L (1 μL Kan stock/mL) at 28°C and 200 rpm. *A. tumefaciens* will grow most efficiently if you make enough that you can dilute ca. 1:10 into your final culture.

Table 1
Guidelines for calculating the amounts of *A. tumefaciens* cultures required for VIGS experiments

	Construct for each gene of interest			pBINTRA			
Plants per construct	Inoculation solution (mL)	YEP culture (mL)	Pre-culture (mL)	Inoculation solution (mL)	YEP culture (mL)	Pre-culture (mL)	Final inoculation solution (mL)
×	0.5×	2.5×	0.25×	0.5×	2.5×	0.25×	×
50	25	125	12.5	25	125	12.5	25 + 25 = 50

The calculation is adjusted to a skilled user and final volume of 1 mL of inoculation culture to be used per plant

6. Water your plants the morning of the day you want to VIGS, or the day before. You will not be able to water them for 2 days after VIGS inoculation.
7. On day 3 after potting (or later depending on plant growth), add 1 mL/mL Kan stock to your remaining prepared Erlenmeyer flasks (or add 1 mL/mL Kan stock to your YEP media and pour into autoclaved flasks). Pour in your pre-cultures (final dilution 1:10). These will need to grow at least 5–6 h to reach the correct OD. Alternatively, you can inoculate a 1:1,000 aliquot of your pre-culture into the prepared Erlenmeyer flasks and grow the cultures overnight (about 16 h) at 28°C and 200 rpm.
8. When the culture reaches an OD600nm 0.4–0.6, centrifuge at 3,000–4,000 × g for 5–10 min. You can centrifuge 40 mL at a time in 50 mL Falcon tubes.
9. Discard the supernatant and resuspend the pellet carefully, without grinding or vortexing, in the inoculation solution of 1:1 10 mM MES:10 mM MgCl2 (final concentration 5 mM MgCl2 and 5 mM MES). The volume of the inoculation solution should be 1/5 the volume of your culture, so that the final OD_{600nm} of the inoculation solution is 2.0–3.0.
10. For every construct, mix a final inoculation solution of half pTVgoi construct and half pBINTRA. To avoid contamination, prepare Falcon tubes with pBINTRA for each pTVgoi construct, then add construct inoculation solutions one by one.
11. Perform inoculations using a 1 mL syringe without a needle: pressure-inject inoculation solution into 3 leaves per plant, with 1–4 inoculations on the underside of each leaf. You should try to saturate at least 75% of each inoculated leaf. It is easiest to inoculate plants during the light period when the stomata are open.
12. Cover the whole tray of plants with a plastic bag or with an upside-down tray (make sure the upside-down tray is tightly closed against the tray holding the plants to provide high humidity for establishment of A. tumefaciens transformation), and leave the lights off for 2 days.
13. Remove the plastic bag or tray and switch on the lights.
14. The viral spread is always accompanied by a characteristic leaf phenotype: after about 8–10 days, newly developing leaves are wrinkled, and leaves which mature post-inoculation often appear lighter than the oldest rosette leaves which were not VIGSed. You should discard plants that do not develop wrinkled leaves by day 14 post-inoculation

as they will not show gene silencing. VIGSed plants should also be smaller than non-VIGSed plants grown under the same conditions. The wrinkled-leaf phenotype will lessen as plants start to elongate.

15. Bleaching of the leaves on the PDS positive control should appear at approximately 10–14 days post-inoculation. You can start your experiments as soon as the leaf positions you want to use have bleached in the PDS control plants (see Subheading 3.4).

3.3.2. Modification of Basic Inoculation Protocol for S. nigrum

While direct infiltration of leaves with bacterial solution in a syringe works well to initiate VIGS in tobacco, *S. nigrum* plants require earlier inoculation at the seedling stage. We therefore developed a method for vacuum-mediated infiltration of young seedlings to improve VIGS efficiency in this plant species. Because the enhanced VIGS system developed by Liu and colleagues (15) showed better efficiency in *S. nigrum*, the use of the vector system pTRV1/pTRV2 is described.

(a) Germination of *S. nigrum* seeds:

1. Prepare fresh sterilization solution with 2% (w/v) of dichlor-oisocyanuric acid sodium salt and 0.005% Tween-20 in dH_2O.
2. Incubate the seeds in the sterilization solution for 5 min, shaking occasionally. The following steps should be carried out under a sterile hood.
3. Wash seeds four times with sterile dH_2O.
4. Stratify seeds in 2 mL 1 M KNO_3 overnight (or up to 3 days) in a fridge at 4°C.
5. Wash with sterile dH_2O and distribute 30–40 seeds on cultivation media using a forceps.
6. Transfer plates to a growth chamber (16 h light/8 h dark, 26°C, 155 μmol s^{-1} m^{-1} PAR at shelf height).

(b) Inoculation of *S. nigrum* seedlings by vacuum infiltration:

1. Four days before plant inoculation, streak *Agrobacterium* cultures on LB agar plates with 50 mg/L kanamycin. Allow to grow for 2–3 days at 28°C. Plates can be stored at 4°C for approximately 1 month.
2. Two days before plant inoculation, inoculate 5 mL LB-medium with a single colony for each construct. Grow overnight at 28°C and 200 rpm.
3. One day before plant inoculation, in the evening: inoculate 200 mL LB-MES with 300 μL culture and grow overnight at 28°C and 200 rpm.

4. When the culture reaches an OD_{600nm} of 0.4–0.8, centrifuge at 2,000 × *g* (at 4°C) for 10 min.
5. Discard the supernatant and resuspend the pellet in 20 mL 1× infiltration medium (per 200 mL original medium). Mix gently until the pellet is completely dissolved.
6. Let stand for 3 h in the dark (or shaded) at RT.
7. Mix the 20 mL of each construct of interest with 20 mL of pTRV1 and fill this solution in 100 mL glass beakers.
8. Carefully transfer whole seedlings (7–10 days after sowing) from the petri dish in the solution; remove large agar clumps if necessary. Place the beakers in the desiccator and apply vacuum until you reach 60–100 mbar. Release vacuum very slowly.
9. Transfer seedlings into pots and water them with few mL of water.
10. Cover the pots with plastic covers and transfer them to a growth chamber (21–22°C, 16/8 h day/night), but keep direct lights switched off during the next day.
11. Remove the cover and switch on the lights on the second day after inoculation.
12. Bleaching of the leaves should start after 10–15 days. Plants reach a reasonable size for experiments after 21 days.

3.4. Screening of Gene Function in VIGS Plants

All experiments with VIGS plants should be conducted in a climate growth chamber with the following light and temperature regime: photoperiod 16 h light/8 h darkness, temperature 20–22°C (22°C is preferable for caterpillar performance; see Note 13), ca. 65% relative humidity and low light intensity (115–165 μmol s^{-1} m^{-2} PAR). Experiments should begin once clear and even bleaching is visible in pTVPD positive control plants; tissues which were already established at the time of inoculation will not bleach, and the corresponding tissues should be avoided in experiments with *goi*-silenced and EV control plants.

3.4.1. Plant Elicitations

(a) Preparation of Lepidopteran larvae oral secretions:
For the collection of *M. sexta* and *S. littoralis* oral secretions and regurgitates (OS), larvae should be reared on WT plants of the same species to be treated (*N. attenuata* or *S. nigrum*) until the third to fifth instars. OS are collected on ice with Teflon tubing into a GC vial with a septum lid connected to a vacuum, and stored under argon at −20°C. OS can be gathered from the larvae twice a day and should be pooled to reduce variability due to differences among collections. Prior to plant treatment, centrifuge OS 10 min at 16,100 × *g* at 4°C to remove plant material, and transfer the supernatant into a

new tube. Active components of *M. sexta* OS have been shown to elicit plant responses in dilutions even up to 1:1,000 (20). To obtain higher volumes of OS solution, OS can be diluted 1/5 (v/v) with dH_2O and stored in ex. 0.5–1 mL aliquots under argon at −20°C.

(b) Elicitation of the leaves with wounding and oral secretions (=simulated herbivory):
For simulated herbivory treatment, leaves are wounded with a pattern wheel and 20 μL of 1/5 diluted OS are immediately rubbed with a gloved finger onto each wounded leaf (W + OS); for wounding treatment, leaves are wounded with a pattern wheel, and 20 μL of dH_2O are rubbed with a gloved finger onto each leaf (W + W).

(c) Elicitation of plants with the plant hormone methyl jasmonate:
For methyl jasmonate (MeJA) treatments, MeJA is dissolved in melted lanolin (50°C) at a concentration of 7.5 mg mL^{-1}. This can be made in larger batches and stored in 1 mL plastic syringes for application (store wrapped in parafilm at −20°C); do not heat lanolin above 50°C to dissolve MeJA, and remove from heat as soon as solution is mixed. Twenty μL of the paste (150 μg MeJA) are applied to the base of a leaf using the 1 mL syringe or a small spatula; 20 μL of pure lanolin are applied to controls.

3.4.2. Determination of Silencing Efficiency by qRT-PCR (See Note 14)

1. Excise samples from EV and *goi*-silenced plants and immediately freeze them in liquid nitrogen. Store samples at −80°C until analysis. At least 3 biologically replicated samples should be collected.
2. Grind samples in liquid nitrogen and aliquot ca. 100 mg of each sample to a fresh 2 mL microcentrifuge tube.
3. Purify total RNA using an RNA mini isolation kit or TRIzol Reagent (Invitrogen). Usually, 100 mg of leaf tissue yields about 30–70 μg of total RNA. Dilute an aliquot of each RNA sample with RNAse-free water to a final concentration 0.5 μg/μL.
4. Prepare cDNA by reverse transcribing 0.5 μg of total RNA using oligo(dT) or random hexamer primers and an RNase H^- reverse transcriptase enzyme.
5. Prepare external standard samples. Make a serial dilution of a cDNA sample which has the highest expected transcript levels of the target gene (dilution factor 2–5, depending on the transcript abundance of gene in the samples that have the lowest transcript levels, e.g., considering a 90% silencing efficiency). Usually 4–6 serially diluted samples are sufficient.
6. Run samples on a qRT-PCR instrument using target gene-specific primers and primers for an internal standard (see Note 15).

The external standard samples should be included on the same plate for comparison.

7. Construct standard curves using the Ct values and log (designated concentrations) of these external standards (see Note 16).
8. Extrapolate the relative concentrations of the target and internal standard genes using the constructed standard curves. Calculate the relative ratios between the concentrations of the genes of interest and those of their respective internal standards.
9. The silencing efficiencies can be obtained by calculating the ratios of the transcript levels of the target genes in *goi*-silenced plants and in EV.

3.4.3. Extraction and Phytohormone Analysis by LC-MS/MS (see Note 17)

1. Harvest plant tissues after the treatments of interest and freeze samples immediately in liquid nitrogen to prevent the modification of the amounts or integrity of the phytohormones. At least 3–5 biologically replicated samples should be collected.
2. Homogenize plant tissue (0.2–0.3 g fresh weight) to a fine powder over liquid nitrogen (homogenization can be done manually with a mortar and pestle; however the use of mechanical grinders facilitates the extraction of a large number of samples) (see Note 18). Aliquot ca. 100 mg frozen material to 2 mL microcentrifuge tubes (record fresh mass of the samples).
3. Add 1 mL of ethyl acetate spiked with 0.1 μg ($^{2}H_{2}$)JA, ($^{2}H_{4}$)SA, and ($^{2}H_{6}$)ABA and 0.05 μg ($^{13}C_{6}$)JA-Ile to the samples, vortex the samples for 10 min at 4°C, and finally centrifuge for 15 min at 13,200*g* (4°C).
4. Transfer the upper organic phase to a fresh 2 mL microcentrifuge tube and re-extract the leaf material/water phase with 0.5 mL ethyl acetate (without labeled IS).
5. After vortexing and centrifugation (step 3), remove the upper organic phase and pool with the first extraction (step 4) to have a final 1.5 mL volume. Evaporate the solvent to dryness without heating under reduced pressure (e.g., Speed-Vac with trap for organic solvents).
6. Reconstitute evaporated samples in 0.4 mL of 70/30 (v/v) methanol–water and centrifuge at 13,200×*g* at 4°C, 15 min. Transfer supernatants to HPLC vials for analysis.
7. Analysis is carried out in a LC-(ESI)-MS/MS instrument. Example for a Varian 1200 Triple-Quadrupole-LC-MS system (Varian, Palo Alto, CA): 0.01 mL of the sample volume is injected in a ProntoSIL® column (C18-ace-EPS, 50×2 mm, 5 μm, 120 Å, Bischoff, Leonberg, Germany) connected to a pre-column (C18, 4×2 mm, Phenomenex, Torrance, CA). As mobile

Table 2
LC-MS parameters use for detection and quantification of phytohormones and IS using Varian 1200 Triple-Quadrupole-LC-MS system

Phytohormone	Capillary E (V)	M1 (*m/z*)	M2 (*m/z*)	Collision E (V)	Rt (min)
SA	35	137	93	15	6
[$^2H_4$] SA	35	141	97	15	6
JA	35	209	59	12	5.9
[$^2H_2$]JA	35	213	59	12	6.6
ABA	35	263	153	9	5.6
[$^2H_6$]ABA	35	269	159	9	5.6
JA-Ile	45	322	130	19	7.3
[$^{13}C_6$]JA-Ile	45	328	136	19	7.3

phases 0.05% formic acid in water (solvent A) and methanol (solvent B) are used in a gradient mode with the following conditions: time/concentration (min/%) for B: 0.0/15; 2.5/15; 4.5/98; 10.5/98; 12.0/15; 15.0/15; time/flow (min/mL): 0.0/0.4; 1.5/0.2; 1.5/0.2; 10.5/0.4; 15.0/0.4. The analytes are detected in the ESI negative mode and multiple reaction monitoring (MRM) according to the parameters shown in Table 2. Additional parameters include: collision gas (2.1 mTorr); API drying gas (19 psi; 300°C); API nebulizing gas (60 psi); needle (4,500 V); shield (600 V); detector (1,800 V).

3.4.4. Determination of Secondary Metabolites by High-Performance Liquid Chromatography

1. Harvest plant tissues after the treatments and freeze immediately in liquid nitrogen to prevent modification or degradation of secondary metabolites. At least 3–5 biologically replicated samples should be collected.
2. Homogenize plant tissue (ca. 0.5 g fresh weight) to a fine powder over liquid nitrogen and aliquot 100–150 mg of the tissue to clean 2 mL microcentrifuge tubes; do not allow samples to thaw. Homogenization can be done manually with a mortar and pestle; however the use of mechanical grinders facilitates the extraction of large number of samples (see Note 18). Record the fresh masses of the samples.
3. Add 10 μL of extraction solvent per each mg of sample; extract samples by vortexing for 10 min at 4°C and centrifuge for 20 min at 13,200×*g* (4°C).
4. Transfer supernatant to a clean 1.5 mL tube and spin again to remove all remaining particles.

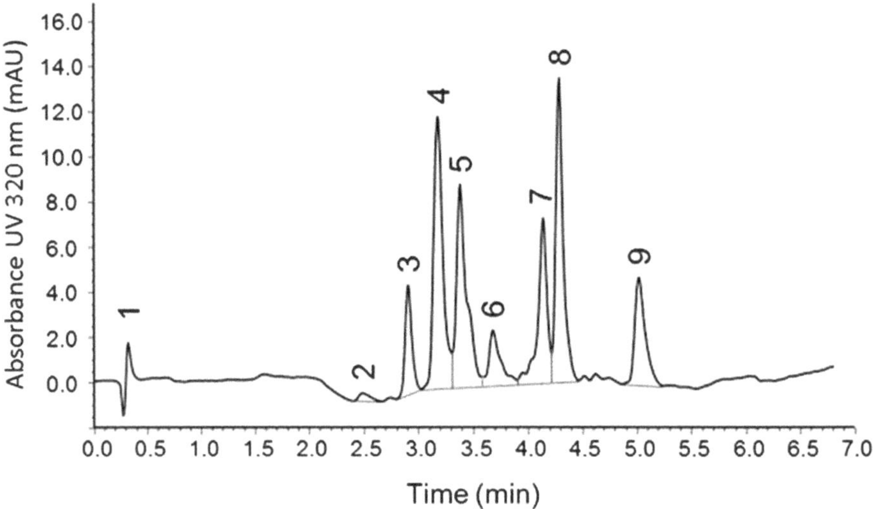

Fig. 3. Typical herbivory-elicited secondary metabolite profile in an *N. attenuata* leaf after feeding by *M. sexta* neonate larva for 4 days; peaks: (1) solvent peak; (2) unknown phenolic (2.5 min); (3) caffeoylputrescine (2.9 min); (4) chlorogenic acid (3.2 min); (5) cryptochlorogenic acid (3.4 min); (6) unknown compound; (7) dicaffeoylsperimidine isomer (4.1 min); (8) dicaffeoylsperimidine isomer (4.3 min); (9) rutin (5.0 min). The nicotine peak at 0.75 min is only visible at 260–260 nm due to its narrow UV absorbance maximum.

5. Transfer ca. 0.5 mL of the supernatant to a clean 1.5 mL tube or appropriate glass vial, if the HPLC is equipped with an auto-injector.
6. Inject 1–10 μL onto an HPLC column to separate the samples. Example for an Agilent 1100 series HPLC (Agilent Technologies, Santa Clara, USA): 1 μL sample is injected in a Chromolith FastGradient RP 18-e column (50 × 2 mm; monolithic silica with bimodal pore structure, macropores with 1.6 μm diameter, Merck, Darmstadt, Germany) attached to a pre-column (Gemini NX RP18, 2 × 4.6 mm, 3 μm). The mobile phases (0.1% formic acid + 0.1% ammonium water, pH 3.5) as solvent A and methanol as solvent B are used in a gradient mode with the following conditions: time/concentration (min/%) for B: 0.0/0; 0.5/0; 6.5/80; 9.5/80; reconditioning 5 min to 0% B. The flow rate is 0.8 mL/min and the column oven temperature is 40°C. A typical chromatogram of a separated *N. attenuata* leaf sample is shown in Fig. 3.
7. Concentrations of secondary metabolites (μg/g FM) are calculated from the external calibration curves detected at the maximum absorbance of each compound (nicotine, 254 nm; chlorogenic acid, 320 nm; rutin, 360 nm). Two abundant phenolamides in tobacco–caffeoylputrescine and dicaffeoylspermidine–can be detected at 320 nm, and concentrations calculated using chlorogenic acid curves (expressed as CGA equivalents).

3.4.5. Bioassays with Specialist and Generalist Herbivores (see Note 19)

The growth chamber temperature should be adjusted to 22°C for caterpillar performance at least 1 day before the experiment (see Note 13). The minimum recommended number of replicates (one plant with one caterpillar) is at least 20 for each VIGS vector type (EV, *goi*).

(a) Procedure for larvae of a specialist herbivore (*M. sexta*) (see Note 20):

1. Arrange the plants so that their leaves do not touch each other to prevent caterpillar movement between plants. Allow extra space for the duration of the experiment and growth of the plants over the course of the bioassay.
2. Select one VIGS-silenced leaf on each plant (leaf position corresponding to bleached tissues in PDS positive controls) and place two neonates on the leaf (see Notes 21, 22). The next morning, remove one neonate from the leaf to keep a single caterpillar on each plant.
3. Weigh caterpillars after 4 days and then every second day (or according to desired experimental design) (see Note 23). After the first weighing at 4 days, consumed leaf area can be determined by taking pictures from a standardized perspective with a size marker or, better, cutting off leaves and scanning them, and analysis with appropriate image analysis software (e.g., ImageJ freeware available from the United States National Institutes of Health). The local herbivore-fed leaf or systemic leaves can be harvested and analyzed for feeding-induced secondary metabolites, phytohormones or gene expression.

(b) Procedure for generalist herbivore larvae *(S. littoralis*):

1. Rear hatched neonates on artificial diet until the fresh mass of the caterpillars reaches approximately 5–10 mg (see Note 24).
2. One day before the start of the experiment, transfer caterpillars into a small plastic box with a perforated lid supplied with 1–2 leaves from individual VIGS-silenced genotypes (EV and goi). This will prevent the transfer of any artificial diet-derived compounds from the larvae to the plant, and allow larvae the transition to plant feeding.
3. Tighten the lower part of the plastic bag around the base of the plant (Fig. 4): make sure that only VIGS-silenced leaves are included (leaf positions corresponding to bleached tissues in PDS positive controls) (see Note 25).
4. Weigh the caterpillars to record the starting mass of each, and place one larva on each plant.

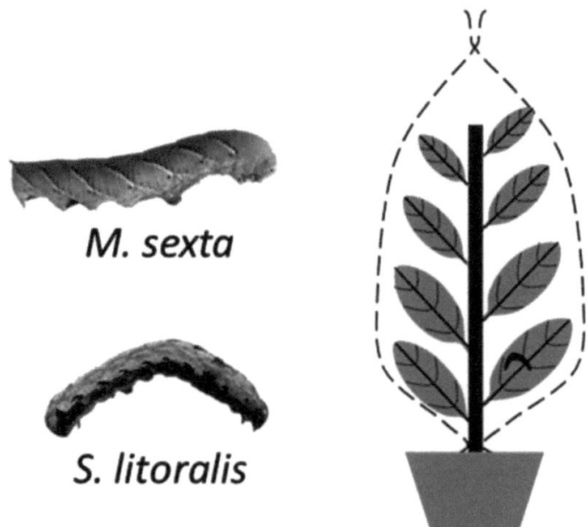

Fig. 4. M. sexta and S. litoralis larvae and herbivory bioassay setup for highly mobile larvae such as Spodoptera spp. A perforated plastic bag is placed and secured around the plant to prevent the movement of larvae from the plant. The size of the bag should be adjusted to the expected growth of the plants during the bioassay (up to 14 days).

5. Roll-up and close the top of the bag with string or soft wire, without damaging the plant. Perforation size should allow gas exchange but should not allow larvae to escape.
6. Weigh caterpillars every 2 days according to the desired experimental design (see Note 26). As for bioassays with specialists, consumed leaf area can be determined by taking pictures from a standardized perspective with a size marker or, better, cutting off leaves and scanning them, and analysis with appropriate image analysis software (e.g., ImageJ freeware available from the United States National Institutes of Health). The local herbivore-fed leaf or systemic leaves can be harvested and analyzed for feeding-induced secondary metabolites, phytohormones, or gene expression.

4. Notes

1. The two commonly used TRV-vector systems (1, 15) exhibit very different silencing efficiencies and degrees of viral symptoms in different plant species (14, 21). We advise assessing which vector yields the most efficient silencing with the least viral symptoms for each new host species. Notably, the vector developed by Liu and colleagues (15), which was designed to enhance viral replication and consequently silencing, requires special control vectors. Plants transfected with an empty control vector of this vector system generally show more severe virus

symptoms when compared to plants inoculated with vectors carrying fragments of genes of interest (14, 22). The use of a control vector which carries a noncoding sequence similar in size to the gene fragments used for silencing solves this problem reliably, and thus allows better and less confounded comparisons between control plants and plants silenced for specific genes.

2. The sequences of pBINTRA and pTV00 vectors can be downloaded from the following link (http://www.plantsci.cam.ac.uk/research/baulcombe/sequencedata.html). The vectors can amplify under kanamycin selection both in *E. coli* and *A. tumefaciens.*

3. Based on our experience, *goi* fragments at a size of about 300 bp provide effective gene silencing; fragments shorter than 150 bp may result in a weak silencing effect, while fragments longer than 300 bp increase the probability of unspecific silencing of genes with short sequence homologies. Note that 23 nucleotides of complete identity may suffice to induce direct post-transcriptional silencing of a gene (23).

4. In a VIGS system, plants are infected with viruses for propagating and activating gene silencing. However, virus infection is known to change hormone and metabolic homeostasis in plant cells. This needs to be always considered when using and interpreting any VIGS result, as hormone crosstalk generally occurs in plants and virus-infected plants will, in most cases, contain higher levels of the pathogenesis-associated hormone salicylic acid. Therefore, in each VIGS experiment, both negative and positive control treatments are essential for proper comparison and interpretation of data.

5. *N. attenuata* seeds can be obtained upon request from Ian T. Baldwin (baldwin@ice.mpg.de), Department of Molecular Ecology, Max Planck Institute for Chemical Ecology in Jena (http://www.ice.mpg.de/ext/).

6. The protocol was optimized for the *S. nigrum* genotype Sn30, an inbred line derived from a wild collection on a field plot near Jena in Germany (24). Seeds can be obtained upon request from Ian T. Baldwin (baldwin@ice.mpg.de), Department of Molecular Ecology, Max Planck Institute for Chemical Ecology in Jena (http://www.ice.mpg.de/ext/).

7. Standard cDNA can be prepared from any total RNA which contains a high amount of target gene transcripts. Generally, cDNA from induced EV plants should be used as it will contain more transcripts than *goi*-silenced tissues.

8. It is recommended to use a photodiode array (PDA) detector for analysis of secondary metabolites, which allows the monitoring of multiple wavelengths in each run. We collect data

from 254 nm for nicotinic alkaloids, 320 nm for phenylpropanoids, and 360 nm for flavonoid (rutin) quantification.

9. We only describe here the cloning procedure for the pTV00 vector designed by Ratcliff and colleagues (1). Detailed information about vectors developed by Liu and colleagues (15) and used for *S. nigrum* VIGS can be obtained from the original publications, (14) and (15).

10. Vector pTV00 carries the complete lac promoter from pUC19, oriented in the same direction as P35S between RB and P35S. To avoid a potential toxic effect from the transcription of the TRV cDNA and the *goi* fragment, the lac promoter should be repressed.

11. *N. attenuata* seeds require smoke cues to break dormancy and germinate. Treatment with smoke solution is not required for other tobacco species like *N. benthamiana*; however, it is still recommended to use the GA_3 solution for increased and synchronized germination in other tobacco species.

12. Proper plant age is critical for the establishment of VIGS. Older plants are usually more resistant to *A. tumefaciens* infection, while younger plantlets may not survive damage to the leaves caused during infiltration. Plants that we normally use for infiltration have leaf blades 2–3 cm long.

13. Low temperature is important for proper VIGS establishment, but the temperature should permit normal development of larvae. We find the best compromise is to establish VIGS at 20°C and to increase the temperature to 22°C one day before placing the caterpillars on the leaves.

14. Compared with many other tools for gene transcriptional analyses, such as northern blotting and semiquantitative reverse transcription-PCR, qRT-PCR has superb specificity, speed, and sensitivity (25, 26). Because silencing efficiencies may range from 50 to 95%, high accuracy of the method is required, allowing detection of at least twofold reduction in transcript levels.

15. Please follow the primer design schemes recommended by the provider of the qRT-PCR instrument. Generally, primers should have 18–26 nucleotides, GC contents of 40–60%, and not more than 3 G or C in the last five nucleotides. The theoretical melting temperatures (salt-adjusted) should be 58–65°C. The amplicon sizes should range from 70 to 200 bp and importantly, the amplicons should not be within the region of the target gene that was used for cloning into the VIGS construct. At least two pairs of primers must be synthesized: one pair for the *goi* and another pair as an internal standard to normalize each sample based on cDNA concentration and quality. These genes are usually housekeeping genes, such as actin, elongation factor, ubiquitin, etc.

16. The slopes of the calibration curves (log base 10 of the relative cDNA concentration versus Ct value) should be around −3.32. Both the linearity and the slope values indicate whether the qRT-PCR and amplification of the targets are successful.
17. The analysis of phytohormones can be achieved by different methods. Among the most commonly used methods are separation by gas or liquid chromatography (GC or LC, respectively) and detection by mass spectrometry (MS). Analysis by GC normally requires partial purification of the phytohormones to be analyzed, and their modification by derivatization to render them less polar and more volatile for GC separation. The main advantage of the method described here, using LC separation and MS detection, is that the samples are analyzed without prior purification and derivatization, avoiding artifacts and loss associated with these processes. Moreover, due to the simplicity of sample preparation, the method can be easily implemented for high-throughput analyses of diverse metabolites including phytohormones. Quantification is based on the use of known amounts of isotopically labeled internal standards (IS) which can be distinguished from the endogenous phytohormones by their differences in mass/charge (*m/z*).
18. Optional procedure for sample grinding using a ball mill instrument: 0.3 g of frozen material can be homogenized in 2 mL microcentrifuge tubes containing two steel beads (ASK, Korntal-Muenchingen, Germany) by shaking the tubes in a Genogrinder (SPEX Certi Prep, Metuchen, NJ) for 30 s at 200–300 strokes min^{-1}. It is important to ensure that tissue is thoroughly ground to a powder prior to extraction; this may require multiple rounds of Genogrinder shaking (with samples placed back to liquid nitrogen in between) or additional grinding by hand.
19. Herbivore bioassays are powerful tools in plant–herbivore interaction research which allow one to identify potential ecological and physiological roles of a defense gene. However, the phenotypes from bioassays are not only depended on the plant genotype (*goi* silencing) but also strongly dependent on environmental factors, which can alter plant nutritional value during the experiment. For this reason, setting up the right and stable conditions for the bioassay is very important for proper interpretation of results.
20. Usually it is possible to transfer neonates of specialist herbivores like *M. sexta* directly to the leaf; however generalist herbivores like *S. littoralis* need to be reared for a limited time on artificial diet before placing the larvae on leaves of *N. attenuata* to ensure survival of the larvae.

21. Because hatching of caterpillars is not synchronous, some early-emerged larvae can be exposed to prolonged starvation and, as a result, will not survive on the plants. We place two caterpillars and later remove one of them to prevent the loss of valuable VIGS plants in the experiment. In our experience, the presence of two neonates for a short time on the leaf (6–12 h) does not influence the growth or performance of the remaining caterpillars.
22. Silenced leaves can be identified by comparing the progression of silencing in PDS-VIGS positive control plants (leaves in comparable positions to bleached PDS-VIGS plants).
23. Gently hold the protruding "horn" of *M. sexta* larvae with a soft pair of forceps; it is sometimes necessary for older larvae to loosen their grip on the plant by gently moving one half of the soft forceps under their prolegs. After weighing of larvae, we usually place them on a new undamaged leaf. Caterpillar-fed and systemic leaves can be collected for analysis of secondary metabolites, phytohormones or gene expression analysis.
24. Generalist larvae may not survive feeding as neonates on plants.
25. Gently hold the caterpillar's body with a forceps or by hand when moving; try to prevent damaging plants as much as possible when tightening the plastic bags. Mechanical damage can induce jasmonic acid levels in plants!
26. In general, depending on the effect of the gene, the differences in performance of caterpillars usually appear after 6 days or later. The bioassays should be terminated before 14 days since most of the leaves will be consumed and the larvae may be preparing for pupation.

References

1. Ratcliff F, Martin-Hernandez AM, Baulcombe DC (2001) Tobacco rattle virus as a vector for analysis of gene function by silencing. Plant J 25:237–245
2. Kessler A, Halitschke R, Baldwin IT (2004) Silencing the jasmonate cascade: induced plant defenses and insect populations. Science 305:665–668
3. Steppuhn A, Gase K, Krock B et al (2004) Nicotine's defensive function in nature. PLoS Biol 2:e217
4. Zavala JA, Patankar AG, Gase K et al (2004) Constitutive and inducible trypsin proteinase inhibitor production incurs large fitness costs in *Nicotiana attenuata*. Proc Natl Acad Sci USA 101:1607–1612
5. Kang JH, Wang L, Giri A et al (2006) Silencing threonine deaminase and JAR4 in *Nicotiana attenuata* impairs jasmonic acid-isoleucine-mediated defenses against *Manduca sexta*. Plant Cell 18:3303–3320
6. Schwachtje J, Minchin PEH, Jahnke S et al (2006) SNF1-related kinases allow plants to tolerate herbivory by allocating carbon to roots. Proc Natl Acad Sci USA 103:12935–12940
7. Wu J, Hettenhausen C, Meldau S et al (2007) Herbivory rapidly activates MAPK signaling in attacked and unattacked leaf regions but not between leaves of *Nicotiana attenuata*. Plant Cell 19:1096–1122
8. Jassbi AR, Gase K, Hettenhausen C et al (2008) Silencing geranylgeranyldiphosphate synthase in *Nicotiana attenuata* dramatically impairs resistance to tobacco hornworm. Plant Physiol 146:974–986

9. Kessler D, Gase K, Baldwin IT (2008) Field experiments with transformed plants reveal the sense of floral scents. Science 321:1200–1202
10. Skibbe M, Qu N, Gális I et al (2008) Induced plant defenses in the natural environment: *Nicotiana attenuata* WRKY3 and WRKY6 coordinate responses to herbivory. Plant Cell 20:1984–2000
11. Heiling S, Schuman M, Schöttner M et al (2010) Jasmonate and ppHsystemin regulate key malonylation steps in the biosynthesis of 17-hydroxygeranyllinalool diterpene glycosides, an abundant and effective direct defense against herbivores in *Nicotiana attenuata*. Plant Cell 22:273–292
12. Kaur H, Heinzel N, Schöttner M et al (2010) R2R3-NaMYB8 regulates the accumulation of phenylpropanoid-polyamine conjugates, which are essential for local and systemic defense against insect herbivores in *Nicotiana attenuata*. Plant Physiol 152:1731–1747
13. Hartl M, Giri A, Kaur H et al (2010) Serine protease inhibitors specifically defend *Solanum nigrum* against generalist herbivores but do not influence plant growth and development. Plant Cell 22:4158–4175
14. Hartl M, Merker H, Schmidt DD et al (2008) Optimized virus-induced gene silencing in *Solanum nigrum* reveals the defensive function of leucine aminopeptidase against herbivores and the shortcomings of empty vector controls. New Phytol 179:356–365
15. Liu Y, Schiff M, Marathe R et al (2002) Tobacco Rar1, EDS1 and NPR1/NIM1 like genes are required for N-mediated resistance to tobacco mosaic virus. Plant J 30:415–429
16. Ploeg AT, Robinson DJ, Brown DJF (1993) RNA-2 of tobacco rattle virus encodes the determinants of transmissibility by trichodorid vector nematodes. J Gen Virol 74:1463–1466
17. Hamilton RI (1974) Replication of plant viruses. Annu Rev Phytopathol 12:223–245
18. Voinnet O (2005) Induction and suppression of RNA silencing: insights from viral infections. Nat Rev Genet 6:206–U201
19. Saedler R, Baldwin IT (2004) Virus-induced gene silencing of jasmonate-induced direct defences, nicotine and trypsin proteinase-inhibitors in *Nicotiana attenuata*. J Exp Bot 55:151–157
20. Schittko U, Preston CA, Baldwin IT (2000) Eating the evidence? *Manduca sexta* larvae can not disrupt specific jasmonate induction in *Nicotiana attenuata* by rapid consumption. Planta 210:343–346
21. Brigneti G, Martin-Hernandez AM, Jin H et al (2004) Virus-induced gene silencing in *Solanum* species. Plant J 39:264–272
22. Wu C, Jia L, Goggin F (2011) The reliability of virus-induced gene silencing experiments using tobacco rattle virus in tomato is influenced by the size of the vector control. Mol Plant Pathol 12:299–305
23. Thomas CL, Jones L, Baulcombe DC et al (2001) Size constraints for targeting post-transcriptional gene silencing and for RNA-directed methylation in *Nicotiana benthamiana* using a potato virus X vector. Plant J 25:417–425
24. Schmidt DD, Kessler A, Kessler D et al (2004) *Solanum nigrum*: a model ecological expression system and its tools. Mol Ecol 13:981–995
25. Gachon C, Mingam A, Charrier B (2004) Real-time PCR: what relevance to plant studies? J Exp Bot 55:1445–1454
26. Deepak S, Kottapalli K, Rakwal R et al (2007) Real-time PCR: revolutionizing detection and expression analysis of genes. Curr Genomics 8:234–251

Chapter 10

Virus-Aided Gene Expression and Silencing Using TRV for Functional Analysis of Floral Scent-Related Genes

Ben Spitzer-Rimon, Alon Cna'ani, and Alexander Vainstein

Abstract

Flower scent is a composite character determined by a complex mixture of low-molecular-weight volatile molecules. Despite the importance of floral fragrance, our knowledge on factors regulating these pathways remains sketchy. Virus-induced gene silencing (VIGS) and virus-aided gene expression (VAGE) are characterized by a simple inoculation procedure and rapid results as compared to transgenesis, allowing screening and characterization of scent-related genes. Here, we describe methods using TRV as a VIGS/VAGE vector for the characterization of scent-related genes, protein compartmentalization studies, and protein subcellular targeting.

Key words: Virus-aided gene expression (VAGE), Virus-induced gene silencing (VIGS), Functional analysis, Volatile, MYB, Silencing, TRV, Anthocyanin

1. Introduction

Flower aroma plays a key role in the interactions between plants and their environment, as well as in man's consumption of flowers for pleasure and industry. Due to the invisibility and highly variable nature of floral scent, which is determined by a complex mixture of small volatile molecules, its study requires a specialized model system (1–3). Petunia (*Petunia hybrida*) has been useful for characterizing many aspects of floral scent production and emission since it has relatively large flowers and a broad spectrum of scent compounds belonging to diverse biochemical pathways/branches (4, 5). Yet, in this plant, the conventional reverse genetics tools, which are critical for any model system, are very limited because the use of standard antisense/RNAi methods with stably transformed plants is prohibitively labor intensive and time consuming (6).

Annette Becker (ed.), *Virus-Induced Gene Silencing: Methods and Protocols*, Methods in Molecular Biology, vol. 975, DOI 10.1007/978-1-62703-278-0_10,

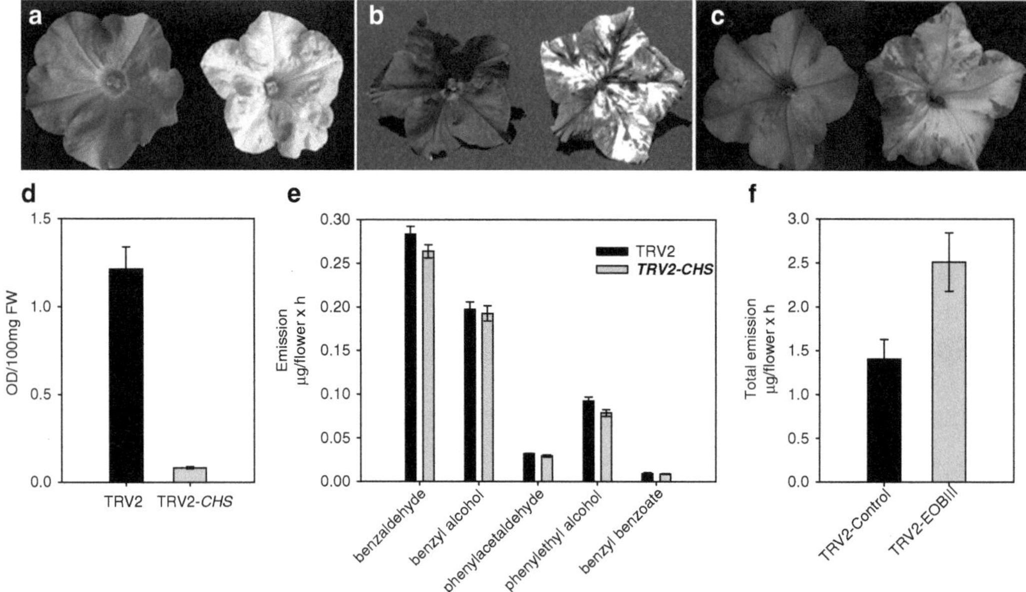

Fig. 1. TRV-based *CHS* silencing as a marker for scent studies. Flowers of petunia lines (**a**) P720, (**b**) P62, and (**c**) B1 infected with *Agrobacterium* carrying pTRV1 and pTRV2-*CHS* (*right*) or pTRV2 (*left*). (**d**) Anthocyanin content in P720 flowers 1 month after infection. (**e**) Levels of major volatiles emitted from corollas of P720 flowers infected with TRV2-*CHS* or TRV2, as determined by dynamic headspace GC-MS analyses. (**f**) Emission of total volatiles in P720 corollas silenced in *EOBIII* using VIGS.

Virus-induced gene silencing (VIGS) and virus-aided gene expression (VAGE) offer alternative approaches for rapid targeted functional analyses characterized by a simple inoculation procedure and applicability to a wide array of plants (7, 8). Among the viruses used to elicit VAGE/VIGS, the TRV (*Tobacco rattle virus*)-based constructs have been most extensively used, producing consistent results for studies of various processes in many plant species, including *Petunia* (8–11). Floral scent-related genes, including transcription factors, have been among the various genes identified and characterized through VAGE/VIGS (9). Since volatile compounds are invisible and VAGE/VIGS are not uniformly spread throughout the infected plant, a visible reporter that points to the modified tissue is employed (9, 12). Suppression of the anthocyanin pathway via *CHALCONE SYNTHASE* (*CHS*) was used to allow easy visual identification of anthocyanin-less silenced flowers with no effect on the level of volatile emissions (Fig. 1a–e) (11). Alternatively, for the VAGE system, fusion of the analyzed gene to or coexpression with easily identifiable marker genes, e.g., genes coding for autofluorescent proteins (AFPs), can be employed (9, 12). The reliability of the VAGE/VIGS system for scent studies is apparent from the consistency of the transient genotypes/phenotypes relative to those characterized through transgenesis. For example, phenotypes of petunia flowers following transient silencing of *BSMT* (encoding SAM benzoic acid/salicylic acid carboxyl methyltransferase), *phenylacetaldehyde synthase*, and *ODORANT1* resulted in changes in volatile organic compounds (VOCs) consistent with the respective genes' functionalities (11).

A transient virus-based system can also be used in a nontargeted screen for novel biosynthetic and regulatory genes involved in floral scent production (9). To this end, a library enriched with floral-specific sequences containing R2R3, bHLH, RWD, WRKY, and EPF domains was generated and cloned into a TRV vector. The TRVs containing individual sequences together with reporter genes were used to infect petunia plants, whose volatile level/composition was then analyzed with the aim of identifying sequences affecting floral scent production. In this way, several novel R2R3-MYB-like regulatory factors of phenylpropanoid volatile biosynthesis, termed *EMISSION OF BENZENOIDS (EOB),* were identified in petunia. For example, transient suppression of *EOBII* expression led to a significant reduction in the levels of volatiles accumulating in and emitted by flowers, such as benzaldehyde, phenylethyl alcohol, benzyl benzoate, and isoeugenol. Upregulation/downregulation of *EOBII* using VAGE/VIGS enabled the identification of several EOBII-target biosynthetic floral scent-related genes encoding enzymes that are directly involved in the production of these volatiles (9) (Fig. 2a–f). VIGS of *EOBIII* (GenBank accession number GQ449251.1, also termed *phMYB4*) (13) enabled its characterization as a negative regulator of floral scent production (Fig. 1f).

Different cellular compartments can participate in the production of volatiles originating from different biosynthetic pathways, e.g., chloroplasts, mitochondria, peroxisomes, and endoplasmic reticulum. Here too, VAGE can be employed for the rapid characterization of intracellular target sites. For example, TRV was used to deliver a gene product to a specific compartment within the cell—the chloroplast (Fig. 2 g–i). This was achieved by infecting petunia with TRV2 carrying the AFP marker gene fused to the transit peptide of *Rubisco small subunit* (TRV2-Rssu:GFP, Fig. 2 g–i) (12). VAGE was also used to confirm subcellular compartmentalization of transcription factors. For instance, the nuclear localization of EOBII was confirmed using TRV2 containing EOBII fused to GFP (EOBII:GFP, Fig. 2j–l).

In this chapter, we describe the methodology and provide protocols for functional analysis of scent-related genes using TRV. Both VAGE and VIGS applications of the TRV-based approach for rapid identification and characterization of genes and their products involved in determining a highly dynamic and invisible trait, floral scent, are presented.

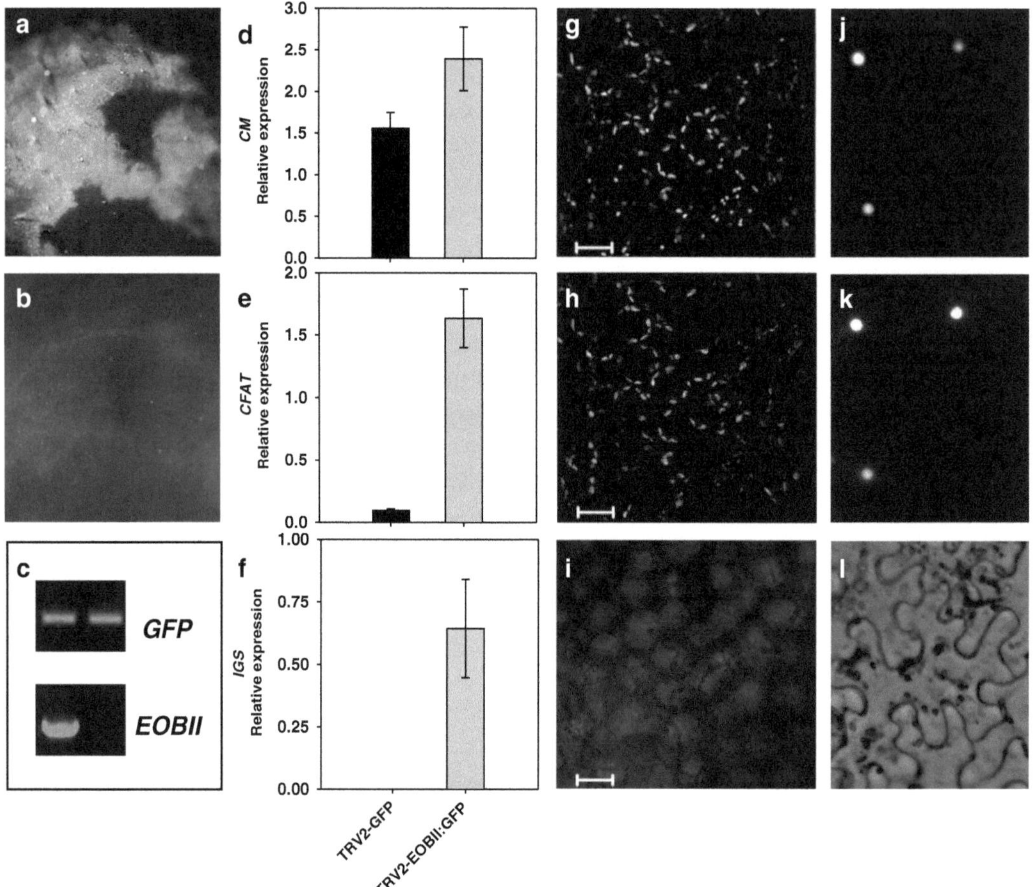

Fig. 2. VAGE of floral scent regulatory factor EOBII in leaves and its subcellular localization. (**a**) GFP fluorescence in *Petunia* leaf tissue transfected with TRV2-EOBII:GFP and (**b**) bright field of the infected tissue. (**c**) RT-PCR analysis of *GFP* and *EOBII* levels in cDNA synthesized from mRNA extracted from petunia leaves infiltrated with *Agrobacterium* carrying TRV2-EOBII:GFP (*left lane*) or TRV2-GFP (*right lane*) as a control. Ethidium bromide-stained agarose gel with the GFP and EOBII products is shown. Quantitative real-time PCR analysis of EOBII target genes: (**d**) chorismate mutase (*CM*), (**e**) coniferyl alcohol acetyltransferase (*CFAT*), and (**f**) isoeugenol synthase (*IGS*) transcript levels in leaves following VAGE of EOBII via TRV2-EOBII:GFP. TRV2-GFP was used as a control. Standard errors indicated by vertical lines. (**g–i**) VAGE of chloroplast-targeted Rssu-GFP reporter gene in leaves of *Petunia* infected with TRV2-Rssu:GFP. (**g**) GFP fluorescence, (**h**) chloroplast autofluorescence, and (**i**) bright field. (**j–l**) Localization of EOBII:GFP fusion protein in leaf epidermis infected with TRV2-EOBII:GFP. (**j**) GFP fluorescence, (**k**) nucleus stained with 4′,6-diamidino-2-phenylindole, and (**l**) bright field.

2. Materials

2.1. Plant Material

1. 3- to 4-week-old rooted petunia plants grown in the greenhouse under 25°C/20°C day/night temperatures and natural photoperiod (see Notes 1 and 2).
2. 3-week-old seedlings of *Nicotiana benthamiana* grown in the greenhouse under 25°C/20°C day/night temperatures and natural photoperiod.

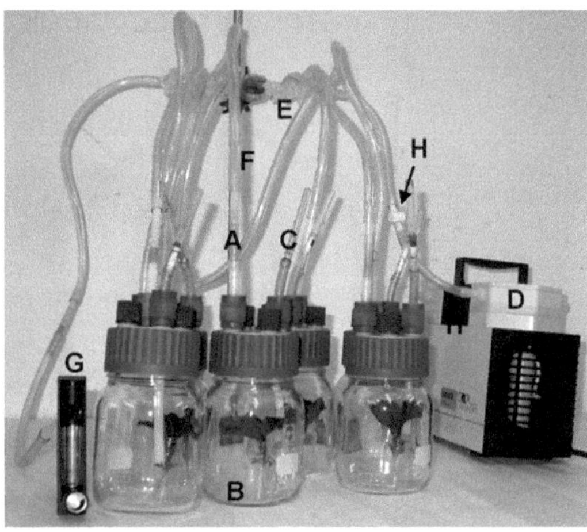

Fig. 3. Floral headspace sampling. (**a**) Outlet trap, (**b**) glass container with floral sample, (**c**) Inlet trap, (**d**) vacuum pump, (**e**) glass splitter, (**f**) tygon tube, (**g**) flow meter, and (**h**) valve.

2.2. Agroinoculation of TRV Vectors

1. *Agrobacterium tumefaciens* (strain AGLO) transformed with pTRV1 and pTRV2 derivatives (9, 11).
2. Luria–Bertani (LB) medium containing 50 mg/L kanamycin and 200 μM 4-hydroxy-3′,5′-dimethoxyacetophenone (acetosyringone) (stock: 200 mM in DDW after dissolving in EtOH).
3. Environmentally controlled (28°C) shaker.
4. Inoculation medium (fresh): 10 mM 2-(*N*-morpholino)ethane sulfonic acid (MES) (stock: 0.5 M, pH 5.5), 200 μM acetosyringone, 10 mM $MgCl_2$ (stock: 1 M), and sterile DDW.
5. Spectrophotometer for measuring optical density.
6. Centrifuge for 15- to 50-mL tubes.
7. 1-mL syringe.

2.3. Dynamic Headspace of Volatile Compounds

1. Growth chamber or growth room with controlled environment.
2. Adsorbent traps consisting of a glass tube containing 100 mg Porapak Type Q (Waters, Milford, MA, USA) mixed with 100 mg 20/40-mesh-activated charcoal (Supelco, Bellefonte, PA, USA) (Fig. 3a).
3. 1-L glass containers, with inlet and outlet, in which to place flowers (Fig. 3a–c).
4. Vacuum pump (Fig. 3d).
5. Glass splitter for multiple samples (Fig. 3e).
6. Tygon tubing to connect the pump to the glass splitter, the splitter to the trap, the trap to the glass container outlet, and another trap to its inlet (Fig. 3f).
7. Flow meter (scaled for 200 mL/h) (Fig. 3g).
8. Valve for flow adjustment (Fig. 3h).

9. *n*-Hexane (Sigma-Aldrich, Rehovot, Israel) for volatile elution.
10. Acetone for cleaning traps.
11. 10 ppm isobutylbenzene (in *n*-hexane) (Sigma-Aldrich) as internal standard.
12. 4-mL vials for trapping eluate (JSi Scientific, Freehold, NJ, USA).
13. Pressurized nitrogen gas cylinder (99.999%, Oxygen and Argon Works, Jerusalem, Israel).
14. 2-mL vial with 200-μL glass insert supplied with spring for GC-MS analyses (JSi Scientific).
15. 20-mL vials (JSi Scientific).
16. Petunia flowers.

2.4. Extraction of Volatiles from Internal Pools

1. Liquid nitrogen.
2. Mortar and pestle.
3. 2- and 15-mL tubes.
4. Shaker in cold room.
5. Refrigerated centrifuge for 15- to 50-mL tubes.
6. Refrigerated centrifuge for 2-mL tubes.
7. *n*-Hexane (Sigma-Aldrich).
8. 2-mL syringe with a 0.2-μM sterile filter.
9. 10 ppm isobutylbenzene (Sigma-Aldrich) as internal standard.
10. 500 mg (fresh weight) petunia corolla.

2.5. GC-MS Analysis of Volatile Compounds

1. A device composed of a Pal autosampler (CTC Analytic, Zwingen, Switzerland), a TRACE GC 2000 equipped with an Rtx-5SIL MS i.d. 0.25 μM, 30 m × 0.25 mm fused silica capillary column (Restek, Bad Homburg, Germany), and a TRACE DSQ quadrupole MS (ThermoFinnigan, Hemel, UK).
2. Helium as the carrier gas.
3. NIST/EPA/NIH Mass Spectral Library (Data Version: NIST 05, Software Version 2.0d) using the XCALIBUR v1.3 program (ThermoFinnigan) for tentative identification of compounds.

2.6. Anthocyanin Extraction

1. Liquid nitrogen.
2. Mortar and pestle.
3. Methanol containing 1% (v/v) HCl.
4. 2- and 15-mL tubes.
5. Refrigerated centrifuge for 15- to 50-mL tubes.
6. Refrigerated centrifuge for 2-mL tubes.
7. Spectrophotometer for measuring optical density.
8. 2-mL syringe with a 0.2-μM sterile nylon filter.
9. 200 mg (fresh weight) corolla tissue.

3. Methods

3.1. Construction of TRV Carrying CHS (VIGS)

1. To generate pTRV2 carrying the petunia *CHS* gene (GenBank accession no. X14599), PCR amplify 250 bp of *CHS* using floral cDNA and the forward primer 5′-*TCCCGGG*GTGGAG GCATTCCAACCATTG-3′ and reverse primer 5′-*GGCCCTC GAG*CATTCAAGACCTTCACCAG-3′ (*Sma*I and *Xho*I sites added during PCR are shown in *italics* in the forward and reverse primers, respectively).
2. Restrict the PCR product using *Sma*I and *Xho*I, and clone the resultant insert into pTRV2 (8–11) restricted with *Sma*I and *Xho*I.

3.2. Construction of TRV Carrying Proteins of Interest Fused to GFP (VAGE)

1. To construct pTRV2 with *EOBII*-fused upstream of GFP (pTRV2-EOBII:GFP), PCR amplify the 591-bp ORF of EOBII using forward primer 5′-AACGAGATGGATAAAAAA CCATGCAACTCTCA-3′ and reverse primer 5′-*TCCTGGT CC*ATCACCATTAAGCAATTGCATGG-3′. The last nine nucleotides at the 5′ end of the reverse primer of EOBII (shown in *italics*) allow generation of a Gly-Pro-Gly amino acid bridge between EOBII and GFP.
2. Ligate (in frame with GFP) into *Hpa*I-restricted pTRV2-GFP (9, 12).

3.3. Agroinoculation of TRV Vectors

1. Plate *A. tumefaciens* carrying pTRV1 and pTRV2 derivatives on LB solid medium supplemented with 50 μg/mL kanamycin and grow for 24 h at 28°C.
2. Use the colonies to prepare *A. tumefaciens* culture in 10 mL LB medium (for each TRV derivative) supplemented with 50 μg/mL kanamycin and 200 μM acetosyringone in a 50-mL tube. Grow overnight at 28°C with shaking at 200 rpm.
3. Harvest the bacteria by centrifugation (3,000 × *g*, 10 min, room temperature).
4. Resuspend the pellet in inoculation buffer to an OD_{600} of 10.
5. Mix the bacteria containing pTRV1 with those containing pTRV2 derivatives in a 1:1 ratio.
6. Incubate the mixture for 3 h at 28°C with shaking at 200 rpm.
7. Apply 200–400 μL of the mixture to the cut surface of the stem, after removing the apical meristems of a 3-week-old petunia plantlet (see Note 3).
8. Grow plants in the greenhouse under 25/20°C day/night temperatures and natural photoperiod till flowering (see Note 4).

3.4. Dynamic Headspace of Volatile Compounds

1. Place detached flowers into a 20-mL vial filled with tap water (each vial will contain three flowers) (see Notes 5 and 6).
2. Place the vials in a 1-L glass container with inlet and outlet.
3. Connect the pump to the glass splitter, the splitter to the trap, the trap to the glass container outlet, and another trap to the container inlet.
4. Set the vacuum pump to a constant air stream of ~200 mL/min for 24 h.
5. To elute trapped volatile compounds, place the traps on top of a 4-mL vial, and elute the volatiles from the trap using 1.5 mL hexane.
6. Add to the eluate 200 μL of internal standard isobutyl benzene (from 10 ppm stock).
7. Concentrate the sample under a nitrogen stream to 100–200 μL.
8. Transfer the sample into 200-μL insert placed inside 2-mL vial.
9. Keep sample at −20°C.
10. Clean the traps using 3 mL acetone and 3 mL hexane before next use.

3.5. Extraction of Volatiles from Internal Pools

1. Collect 0.5 g (fresh weight) petal tissues with reduced anthocyanin content into 15-mL tube and freeze in liquid nitrogen. Sample can be stored at −80°C (*see* **Note 6**).
2. Grind sample in liquid nitrogen.
3. Add 2 mL hexane containing 200 μL isobutyl benzene (from 10 ppm stock) as internal standard for each sample.
4. Incubate 2 h at 4°C with shaking at 150 rpm.
5. Centrifuge the extract (10,500 × *g*, 4°C) for 10 min.
6. Transfer the supernatant into 2-mL tube.
7. Centrifuge (20,000 × *g*, 4°C) for 10 min.
8. Filter the supernatant through 0.2-μM sterile filter using a 2-mL syringe.
9. Concentrate the samples under nitrogen stream to 200 μL.
10. Transfer the sample into 2-mL vial with 200-μL glass insert.
11. Keep samples at −20°C.

3.6. GC-MS Analysis of Volatile Compounds

1. Inject 1 μL sample into GC using the Pal autosampler.
2. Use helium as the carrier gas at a flow rate of 0.9 mL/min.
3. Set the injection temperature to 250°C (splitless mode), the transfer-line temperature to 240°C, and adjust ion source to 200°C.
4. The analysis is performed under the following temperature program: 2 min of isothermal heating at 40°C followed by a 10°C/min oven temperature ramp to 250°C.

5. Equilibrate the system for 1 min at 70°C before injection of the next sample.
6. Mass spectra are recorded at 3.15 scan/s with a scanning range of 40–450 mass-to-charge ratio and an electron energy of 70 eV.
7. Identify the compounds (>95% match) based on NIST/EPA/NIH Mass Spectral Library (Data Version: NIST 05, Software Version 2.0d) using the XCALIBUR v1.3 program.
8. For further identification of major compounds, compare the mass spectra and retention times with those of authentic standards (Sigma-Aldrich) analyzed under similar conditions.

3.7. Anthocyanin Extraction

1. Collect 200 mg (fresh weight) petal tissues into 15-mL tube, and freeze in liquid nitrogen. Sample can be stored at −80°C.
2. Grind the sample in liquid nitrogen.
3. Add 10 mL methanol containing 1% (v/v) HCl.
4. Incubate for 2 h at 4°C with shaking at 150 rpm in the dark.
5. Centrifuge the extract (10,500 × *g*, 4°C) for 10 min.
6. Transfer 2 mL of supernatant into 2-mL tube.
7. Centrifuge (20,000 × *g*, 4°C) for 10 min.
8. Filter the supernatant through 0.2-μM filter using a 2-mL syringe.
9. Measure the absorption values of the extract at A_{530} and A_{657}.
10. Determine anthocyanin content using the formula $A_{530} - 0.25A_{657}$ to subtract chlorophyll interference.

4. Notes

1. Since the efficiency of TRV-driven gene expression/silencing varies among plant varieties (Fig. 10.1a–c), candidate varieties must be characterized for their susceptibility to VAGE/VIGS. It is most convenient to employ reporter genes with easily identifiable activities (e.g., *CHS* for VIGS, or *AFP* for VAGE). For VIGS, pigmented plants are preferable.
2. It is highly recommended that the plantlets used for infection be virus-free, otherwise efficiency of silencing and of overexpression will be reduced.
3. To get sufficient yield for the analysis of transcript levels, headspace and pool levels of volatiles, infect 5–10 plants per treatment.
4. For microscopy analyses, infect the plants by infiltrating the bacterial mixture into the abaxial side of the leaf using a

syringe without a needle. Grow plants after infection for 24–36 h at 22°C.

5. To optimize the analysis of volatiles in the VIGS/VAGE-affected tissue, harvest flowers with significantly reduced anthocyanin content or strong AFP signal.
6. The efficiency of silencing/overexpression (e.g., *CHS*, *GFP*) generally decreases 2 months after infection, and it is therefore recommended that all analyses be performed within this time frame. To quantify the transcript level of the endogenous plant target gene following VIGS, at least one of the primers should be designed according to the sequence outside of that cloned into TRV2. This is so that the sequence present in the viral vector will not be amplified.

Acknowledgments

We thank Danziger Innovations Ltd. for the support and assistance. This work in our laboratory was funded by Israel Science Foundation grant no. 269/09 and 432/10 and BARD grant no. US-4322-10. A.V. is an incumbent of the Wolfson Chair in Floriculture.

References

1. Farhi M, Lavie O, Masci T et al (2010) Identification of rose phenylacetaldehyde synthase by functional complementation in yeast. Plant Mol Biol 72:235–245
2. Pichersky E, Noel JP, Dudareva N (2006) Biosynthesis of plant volatiles: nature's diversity and ingenuity. Science 311:808–811
3. Zuker A, Tzfira T, Ben-Meir H et al (2002) Modification of flower color and fragrance by antisense suppression of the flavanone 3-hydroxylase gene. Mol Breed 9:33–41
4. Maeda H, Shasany AK, Schnepp J et al (2010) RNAi suppression of *arogenate dehydratase1* reveals that phenylalanine is synthesized predominantly via the arogenate pathway in petunia petals. Plant Cell 22:832–849
5. Moyal Ben Zvi M, Negre-Zakharov F, Masci T et al (2008) Interlinking showy traits: co-engineering of scent and colour biosynthesis in flowers. Plant Biotechnol J 6:403–415
6. Gerats T, Strommer J (2009) Petunia: Evolutionary. Developmental and Physiological Genetics, Springer, New York
7. Purkayastha A, Dasgupta I (2009) Virus-induced gene silencing: a versatile tool for discovery of gene functions in plants. Plant Physiol Biochem 47:967–976
8. Burch-Smith TM, Anderson JC, Martin GB et al (2004) Applications and advantages of virus-induced gene silencing for gene function studies in plants. Plant J 39:734–746
9. Spitzer-Rimon B, Marhevka E, Barkai O et al (2010) EOBII, a gene encoding a flower-specific regulator of phenylpropanoid volatiles' biosynthesis in petunia. Plant Cell 22:1961–1976
10. Chen JC, Jiang CZ, Gookin TE et al (2004) Chalcone synthase as a reporter in virus-induced gene silencing studies of flower senescence. Plant Mol Biol 55:521–530
11. Spitzer B, Moyal Ben Zvi M, Ovadis M et al (2007) Reverse genetics of floral scent: application of tobacco rattle virus-based gene silencing in petunia. Plant Physiol 145:1241–1250
12. Marton I, Zuker A, Shklarman E et al (2010) Nontransgenic genome modification in plant cells. Plant Physiol 154:1079–1087
13. Colquhoun TA, Kim JY, Wedde AE et al (2010) PhMYB4 fine-tunes the floral volatile signature of *Petunia hybrida* through PhC4H. J Exp Bot 62:1133–1143

Chapter 11

Virus-Induced Gene Silencing in Soybean and Common Bean

Chunquan Zhang, Steven A. Whitham, and John H. Hill

Abstract

Plant viral vectors are useful for transient gene expression as well as for downregulation of gene expression via virus-induced gene silencing (VIGS). When used in reverse genetics approaches, VIGS offers a convenient way of transforming genomic information into knowledge of gene function. Efforts to develop and improve plant viral vectors have expanded their applications and have led to substantial advances needed to facilitate gene function studies in major row crops. Here, we describe a DNA-based *Bean pod mottle virus* (BPMV) vector system for both gene expression and VIGS in soybean and common bean.

Key words: Plant, Transient, Gene expression, Gene silencing, VIGS, BPMV, Functional genomics, Legume

1. Introduction

Plant virus-based vectors have been developed to express heterologous proteins in plants for the study of gene function, production of pharmaceuticals, analysis of plant–microbe interactions, fungicide and insecticide screening, metabolic engineering, and nutrient improvement. Plant viral gene expression vectors have many advantages over conventional transgenic technology for protein expression. They are relatively fast and low cost and can produce high yields of recombinant protein, and they can be used in a variety of genetic backgrounds. Plant viral vectors also have applications as virus-induced gene silencing (VIGS) tools for reverse and forward genetic studies of gene function (1). VIGS can be used to specifically downregulate a single gene, members of a gene family, or sets of distinct genes (3, 4, 7). Due to these advantages, many positive-sense RNA plant viruses have been developed as vectors for production of recombinant proteins or as VIGS vectors for many plant species (2, 5, 6, 10). With the rapid increase in genomic information, VIGS vectors provide excellent tools that help to advance

Annette Becker (ed.), *Virus-Induced Gene Silencing: Methods and Protocols*, Methods in Molecular Biology, vol. 975, DOI 10.1007/978-1-62703-278-0_11,

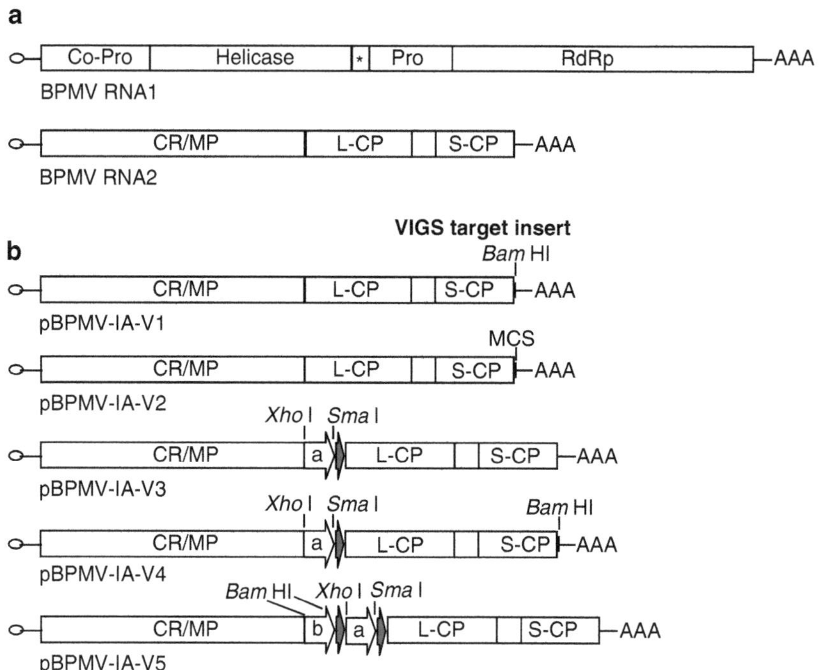

Fig. 1. Schematic representation of the vector set derived from BPMV isolate IA-Di1. (**a**) BPMV genomic RNA1 and RNA2 organization. (**b**) Cloning site for insertion of foreign sequences into BPMV RNA2 derivatives. *Co-Pro* proteinase cofactor; *asterisk* VPg *arrows* gene; *Pro* proteinase; *RdRp* RNA-dependent RNA polymerase; *CR* cofactor for RNA2 replication; *MP* movement protein; *L-CP* large coat protein subunit; *S-CP* small coat protein subunit; *a* and *b* indicate the foreign gene inserts; designates *FMDV-2A* (foot and mouth disease virus 2A proteinase); *MCS* multi-cloning site containing *BamH* I, *Sal* I, *Stu* I, and *Xho* I.

gene function studies in both monocot and dicot plants. Among the different viruses, *Bean pod mottle virus* (BPMV) exhibits unique properties that allow it to be used in soybean for expression of foreign genes and as a VIGS vector to silence genes (8–10). An advantage is its potential for use in marker-assisted VIGS (10).

BPMV, a member of the genus *Comovirus*, has a bipartite positive-sense RNA genome (8, 10). Both RNA1 and RNA2 have a genome-linked viral protein (VPg) covalently linked to the 5′ terminus and a 3′-terminal poly(A) tail. RNA1 (≈ 6 kb) contains a single large open reading frame (ORF) that upon translation produces the polyprotein precursor that is further processed into five mature proteins (from 5′ to 3′: cofactor, helicase, VPg, protease, polymerase) (Fig. 1a). Because BPMV RNA1 encodes all the factors required for genome replication, it can replicate without RNA2. RNA2 (≈ 3.6 kb) codes proteins required for virus movement and virion formation. Depending on which start codon (AUG at positions 467–469 or positions 773–775 for strain IA-Di1) is used, RNA2 can be translated into two forms of polyproteins that are further processed into four mature proteins (from 5′ to 3′: cofactor of RNA2 replication (CR), movement protein (MP), large coat protein (L-CP), and small coat protein (S-CP)) (Fig. 1a).

To develop an efficient BPMV-based viral gene expression and VIGS vector set, cDNA clones of RNA1 and RNA2 were placed under control of a *Cauliflower mosaic virus* (CaMV) 35S promoter and a nopaline synthase (NOS) terminator. In the BPMV RNA2 cDNA construct, restriction sites for foreign gene expression or VIGS were introduced between MP and L-CP or after the stop codon at the end of the S-CP (Fig. 1a). Foreign gene expression is achieved by inserting the coding sequence between the MP and L-CP and in frame with the viral polyprotein. The foreign protein is cleaved from the viral polyprotein by a natural protease cleavage site in MP and addition of an artificial cleavage site between the cloning site and the L-CP. VIGS can be achieved by inserting a fragment of a target gene between the MP and L-CP and in frame with the viral polyprotein or after the S-CP. Insertion after the S-CP provides the advantage that target gene fragments may be inserted in either the sense or antisense orientation and that non-coding sequences may also be targeted for gene silencing.

To initiate infection for gene expression or VIGS, DNAs of BPMV RNA1 and RNA2 constructs are mixed and directly inoculated onto soybean or common bean plants via biolistic delivery. Viral RNA synthesized inside the plant cell following biolistic delivery presumably serves as templates for further replication of viral genomic RNA by the viral-encoded RNA-dependent RNA polymerase. Systemic infection by the recombinant BPMV will then enable foreign gene expression or VIGS of the targeted plant host sequences.

This chapter describes in detail the protocol to perform BPMV-mediated gene expression and VIGS assays in soybean and common bean.

2. Materials

2.1. Foreign Gene Expression and VIGS by Biolistic Delivery Inoculation

1. Soybean seeds.
2. Common bean (*Phaseolus vulgaris*) seeds.
3. Soil mix prepared from equal ratios of Perlite, Sunshine Peat Moss Grower Grade Green, and Metro-Mix 900 Professional Growing Mix (Sun Gro Horticulture Ltd., Vancouver, British Columbia, Canada).
4. Vector pBPMV-IA-R1M (10).
5. Vector pBPMV-IA-R2 (10).
6. Vector pBPMV-GFP (green fluorescence protein) as a gene expression positive control (10).
7. Vector pBPMV-*Gm*PDS (phytoene desaturase) as a VIGS positive control (10).
8. 1.0 micron spherical gold microcarrier particles (Bio-Rad Laboratory, Hercules, CA, USA).

9. Macrocarrier disks, stopping screens, and 1,100 psi rupture disks (Bio-Rad Lab).
10. PDS-1000 biolistic particle delivery system (Bio-Rad Lab).
11. Isopropanol, 70% Ethanol, and 100% Ethanol.
12. All-purpose fertilizer (Peters Excel 15-5-15, Everris NA Inc., Marysville, OH, USA).

2.2. Foreign Gene Expression and VIGS Using Infected Leaf Tissue

1. Soybean seeds.
2. Common bean (*P. vulgaris*) seeds.
3. Fresh or dried leaf tissues infected by pBPMV-IA-R1M and pBPMV-IA-R2.
4. Fresh or dried leaf tissues infected by pBPMV-IA-R1M and pBPMV-GFP as a gene expression positive control.
5. Fresh or dried leaf tissues infected by pBPMV-IA-R1M and pBPMV-*Gm*PDS as a VIGS positive control.
6. BPMV inoculation buffer (pH 7.0 containing 50 mM potassium phosphate).
7. Mortar and pestle.
8. Drierite, carborundum, and cheesecloth.

3. Methods

3.1. Foreign Gene Expression and VIGS by Biolistic Delivery Inoculation

3.1.1. Growing Plants for VIGS

1. Germinate soybean or common bean seeds in soil in pots (1.5 in. (width) × 2.5 in. (length) × 2.5 (depth)). Place the pot in a chamber of 20–25°C with a photoperiod of 16 h (see Note 1). Plants are ready for biolistic delivery inoculation when the primary leaves are fully expanded.

3.1.2. Construction of pBPMV RNA2 Construct for Foreign Gene Expression

1. Select the ORF of a foreign gene for expression and clone into vector pBPMV-IA-V3, pBPMV-IA-V4 or pBPMV-IA-V5 (Fig. 1b) using the *Xho*I and *Sma*I restriction enzymes.
2. Confirm target gene insertion and reading frame by restriction digestion or sequencing using primer BP-R2-1690F (GGT GCTGGTTCACATTCTTC).

3.1.3. Construction of pBPMV RNA2 Containing Target Gene for Silencing

1. Select 300–600 bp region of a target gene for silencing and clone into pBPMV-IA-V1, pBPMV-IA-V2, or pBPMV-IA-V4 (Fig. 1b) using the corresponding restriction enzymes (see Note 3).

2. Confirm target gene fragment insertion by restriction digestion or sequencing using primer BP-R2-3195F (CCTCATTG GTACAAGTGTTT).

3.1.4. Biolistic Inoculation of Soybean or Common Bean

1. Prepare gold microcarriers (see Note 4).
2. Mix 2 μg pBPMV-IA-R1M with 2 μg pBPMV-GFP or gene expression construct plasmid DNA in a volume no more than 40 μL.
3. Mix 2 μg pBPMV-IA-R1M with 2 μg pBPMV-*Gm*PDS or VIGS construct plasmid DNA in a volume no more than 40 μL.
4. Add 25 μL gold microcarrier suspension to the tubes prepared from step 2 or 3.
5. Vortex briefly; to each tube add 85 μL premixed buffer containing 50 μL 50% glycerol, 25 μL 2.5 M $CaCl_2$, and 10 μL 0.1 M spermidine.
6. Continue vortexing for 1 min; centrifuge briefly to pellet the gold. Discard the supernatant and wash pellet once with 70 μL 70% isopropanol, repeat wash once with 70 μL 100% isopropanol, and then resuspend in 30 μL 100% isopropanol.
7. While vortexing, remove 6 μL aliquots of gold and spread on the middle of the macrocarrier; air dry for a few minutes.
8. Assemble gene gun shooting set with stopping screens and the 1,100 psi rupture disks.
9. Place bean seedling inside vacuum chamber and hold the primary leaves flat against a plexiglass support with a metal mesh. Shoot when the vacuum reaches 25–28 in. Hg. Remove the seedling, mist with tap water, and cover for 12–24 h to maintain high humidity.
10. Transfer the inoculated plants into 5 in. diameter pots and maintain plants at 20–25°C in the growth chamber or in greenhouse unless temperature is excessively hot.
11. Fertilize plants after biolistic inoculation at a 10–12 day interval. Dissolve and dilute the all-purpose fertilizer (Peters Excel 15-5-15, Everris NA Inc.) to 36 ppm N. Use appropriately 100 mL of the diluted fertilizer per 5-in. pot.
12. Green fluorescence by pBPMV-GFP can be seen on the primary leaves at about 5 days postbiolistic inoculation under UV light and after about 14 days postbiolistic inoculation on the upper systemic leaves.
13. Bleaching phenotype due to suppression of PDS gene induced by pBPMV-*Gm*PDS should be observed between 16 and 18 days postinoculation (see Note 5).

3.2. Foreign Gene Expression and VIGS Using Infected Leaf Tissues

1. Germinate soybean and common bean seeds in 5 in. diameter pots, maintain and fertilize as described above. Seedlings with two primary leaves are ready for inoculation.

3.2.1. Growing Plants for VIGS

1. Three weeks postbiolistic inoculation, symptoms of infection can be visually observed, and symptomatic leaves are ready for collection and drying.

3.2.2. Preparation of Dry Infected Leaf Tissue for Long-Term Storage

2. Harvest leaf tissues of second trifoliolate as well as the upper youngest leaflets; then lyophilize for 24 h or longer until completely dry.
3. Crush the dry leaves and put into a 50 mL conical tube containing drierite at the bottom.
4. Place the sealed tube at –20°C for long-term storage.

3.2.3. Mechanical Inoculation Using Fresh or Dry Infected Leaf Tissues

1. Grind fresh or dry leaf tissue in KP inoculation buffer using a pestle and mortar on ice (see Note 6). The ratio of leaf tissue (weight in grams) to KP buffer (volume in mL) is 1 to 10–20 for fresh leaf and for 1 to 20–40 for dry tissue.
2. Dust primary leaves of soybean seedlings (or bean leaves) with 600 mesh carborundum.
3. Use a cheesecloth to absorb the sap from the ground tissue and rub inoculate the two primary leaves. Do not rub too hard.
4. Maintain the inoculated plants in the growth chamber as mentioned above, water as needed.
5. Three to five days after inoculation, fertilize once and then maintain fertilization schedule as described in step 11 of Subheading 3.1 paragraph "Biolistic Inoculation of Soybean or Common Bean."
6. Symptoms should appear between 2 and 3 weeks (see Note 8).

4. Notes

1. It is best to germinate seeds in a growth chamber. A greenhouse can be used for this purpose except during hot summer days.
2. High infection efficiency can be achieved when soybean or common bean primary leaves are fully expanded.
3. Silencing efficiency is generally higher for the 3′ region of the target gene ORF. Multiple gene silencing can be achieved by fusing specific VIGS target sequences into one VIGS construct or using a conserved region of multiple genes.

4. Add 0.5 mL 100% ethanol to 30 mg gold in a 1.5 mL tube, vortex well and incubate at room temperature for 10 min, centrifuge briefly to pellet the gold, discard the supernatant and resuspend gold in 0.5 mL H_2O, repeat the water wash twice, and resuspend in 0.5 mL H_2O. While vortexing, aliquot 25 or 50 μL of gold suspension into 1.5 mL tubes and store at −20°C.
5. Always perform pBPMV-IA-R1M with pBPMV-IA-R2 inoculation as a negative control and pBPMV-IA-R1M with pBPMV-*GmPDS* inoculation as a positive VIGS control. Suppression of *PDS* leads to the inhibition of the carotenoid synthesis, causing the plants to exhibit a photobleached phenotype (7). Photobleaching starts on the upper systemic leaves at days 15–18.
6. Successful infection rate using biolistic inoculation in soybean is about 50–80%. Therefore, at least six individual plants should be biolistically inoculated per construct. The mechanical inoculation rate should be close to 100%.
7. The best silencing results are typically on the third and fourth trifoliolates (8–10).
8. Symptom severity of BPMV infection depends on cultivars, BPMV RNA1 used, BPMV RNA2 construct, and environment particularly temperature and fertilization condition. For example, optimal symptom expression occurs on cultivar 'Williams' at a temperature of 20–25°C with a photoperiod of 16 h under fertilization schedule specified at step 11 of Subheading 3.1 paragraph "Biolistic Inoculation of Soybean or Common Bean." Since pBPMV-IA-R1 determines plant phenotype when inoculated with BPMV (10), select pBPMV-IA-R1 that will allow BPMV-VIGS without confounding phenotype associated with loss-of-function phenotype (10).

Acknowledgments

This work was supported by the Iowa State University Plant Sciences Institute Innovation Research Grants Program, the North Central Soybean Research Program, the Iowa Soybean Association, the United Soybean Board, the National Science Foundation Plant Genome Research Program (grant no. 0820642), and Hatch Act and State of Iowa funds.

References

1. Burch-Smith TM, Anderson JC, Martin GB et al (2004) Applications and advantages of virus-induced gene silencing for gene function studies in plants. Plant J 39:734–746
2. Di Stilio VS, Kumar RA, Oddone AM et al (2010) Virus-induced gene silencing as a tool for comparative functional studies in *Thalictrum*. PLoS One 5:e12064

3. Lu R, Malcuit I, Moffett P et al (2003) High throughput virus-induced gene silencing implicates heat shock protein 90 in plant disease resistance. EMBO J 22:5690–5699
4. Peele C, Jordan CV, Muangsan N et al (2001) Silencing of a meristematic gene using geminivirus-derived vectors. Plant J 27: 357–366
5. Purkayasthaa A, Dasgupta I (2009) Virus-induced gene silencing: a versatile tool for discovery of gene functions in plants. Plant Physiol Biochem 47:967–976
6. Tang Y, Wang F, Zhao J et al (2010) Virus-based microRNA expression for gene functional analysis in plants. Plant Physiol 153: 632–641
7. Turnage MA, Muangsan N, Peele CG et al (2002) Geminivirus-based vectors for gene silencing in Arabidopsis. Plant J 30:107–117
8. Zhang C, Ghabrial SA (2006) Development of *Bean pod mottle virus*-based vectors for stable protein expression and sequence-specific virus-induced gene silencing in soybean. Virology 344:401–411
9. Zhang C, Yang C, Whitham SA et al (2009) Development and use of an efficient DNA-based viral gene-silencing vector for soybean. Mol Plant Microbe Interact 22:123–131
10. Zhang C, Bradshaw JD, Whitham SA et al (2010) The development of an efficient multipurpose *Bean pod mottle virus* viral vector set for foreign gene expression and RNA silencing. Plant Physiol 153:52–65

Chapter 12

Functional Genomic Analysis of Cotton Genes with Agrobacterium-Mediated Virus-Induced Gene Silencing

Xiquan Gao and Libo Shan

Abstract

Cotton (*Gossypium* spp.) is one of the most agronomically important crops worldwide for its unique textile fiber production and serving as food and feed stock. Molecular breeding and genetic engineering of useful genes into cotton have emerged as advanced approaches to improve cotton yield, fiber quality, and resistance to various stresses. However, the understanding of gene functions and regulations in cotton is largely hindered by the limited molecular and biochemical tools. Here, we describe the method of an *Agrobacterium* infiltration-based virus-induced gene silencing (VIGS) assay to transiently silence endogenous genes in cotton at 2-week-old seedling stage. The genes of interest could be readily silenced with a consistently high efficiency. To monitor gene silencing efficiency, we have cloned cotton *GrCla1* from *G. raimondii*, a homolog gene of *Arabidopsis Cloroplastos alterados 1* (*AtCla1*) involved in chloroplast development, and inserted into a tobacco rattle virus (TRV) binary vector pYL156. Silencing of *GrCla1* results in albino phenotype on the newly emerging leaves, serving as a visual marker for silencing efficiency. To further explore the possibility of using VIGS assay to reveal the essential genes mediating disease resistance to *Verticillium dahliae*, a fungal pathogen causing severe Verticillium wilt in cotton, we developed a seedling infection assay to inoculate cotton seedlings when the genes of interest are silenced by VIGS. The method we describe here could be further explored for functional genomic analysis of cotton genes involved in development and various biotic and abiotic stresses.

Key words: Cotton, Virus-induced gene silencing, Functional genomics, *Agrobacterium*, *Verticillium dahliae*

1. Introduction

Cotton (*Gossypium spp.*) is widely planted around the world for its significant economic value of the textile fiber, feed, foodstuff, oil, and biofuel products (1). The upland cotton, *G. hirsutum*, dominates world cotton commerce with more than 90% of the annual fiber production (2, 3). As of 2009/2010, cotton cultivation was estimated to be about 30 million hectares worldwide, producing a value of approximately 39 billion US dollars, while its production in

Annette Becker (ed.), *Virus-Induced Gene Silencing: Methods and Protocols*, Methods in Molecular Biology, vol. 975, DOI 10.1007/978-1-62703-278-0_12, © Springer Science+Business Media New York 2013

the United States alone was estimated at approximately five billion dollars (4).

The major concern for cotton production is the significant loss of yield caused by many devastating diseases and pests. Cotton has been considered as the world's *dirtiest* crop due to the heavy application of pesticides and fungicides. The soilborne pathogen *Verticillium dahliae* causes severe *Verticillium* wilt diseases on cotton (5). Because of extremely persistent resting structures, such as microsclerotia, this pathogen can survive in soil for many years. Most notably, this fungus is very difficult to be reached by fungicides because the fungi reside in the woody vascular tissues and can be transmitted systemically in cotton plants (5). Despite the significant efforts towards understanding the biology of this pathogen and identifying the genetic determinants responsible for cotton *Verticillium* resistance (6, 7), to date, the genetic and molecular mechanisms underlying cotton resistance to *Verticillium* infection remain poorly understood.

In recent years, significant advances in cotton genetics and genomics have been achieved towards the molecular breeding and genetic engineering of new cotton varieties to increase the sustainable yield and fiber quality as well as to improve the traits combating various pathogen infections (8, 9). Understanding cotton gene functions and regulations constitutes a critical step for manipulating cotton genes in agriculture. A persistent challenge in cotton functional genomic studies is the lack of molecular and genetic tools partly due to the large genome size, the long growth cycle, and the unstable transformation efficiency (9). Virus-induced gene silencing (VIGS) has been demonstrated as a rapid and efficient approach to study gene functions at whole-genome level in various plant species (10–12). VIGS, a type of RNA-mediated posttranscriptional gene silencing, functions as an antivirus defense mechanism in plants (10–12). Through *Agrobacterium* infiltration, the T-DNA containing the partial viral genome and gene of interest is delivered into host cells. The production of double-stranded RNAs between the endogenous gene and DNA fragment delivered from T-DNA vector results in a chain reaction to generate robust silencing signals (12). With the time, the silencing of endogenous genes occurs both locally and systemically throughout the plant tissues.

To date, different plant virus vectors have been deployed for VIGS assays in dicotyledonous plant species, including *tobacco mosaic virus* (TMV), *potato virus X*, *tomato golden mosaic virus*, *tobacco rattle virus* (TRV), and *cotton leaf crumple virus* (CLCrV) vectors (13–16). In monocotyledonous plants, *barley stripe mosaic virus* has been applied to silence genes in barley and wheat and *brome mosaic virus* (BMV) in rice (17–19). Among these viruses, TRV invades a wide range of hosts and spreads vigorously throughout the entire plants but usually triggers a mild symptom, which makes it a good candidate as a VIGS vector (13). TRV belongs to the Tobravirus containing a bipartite positive-sense single-stranded RNA: RNA1 and RNA2. RNA1 contains genes of the viral replicase, RNA-dependent RNA polymerase, and movement protein, which are required for replication and movement (13). RNA2 contains genes encoding the coat protein and other nonessential structural proteins, which can be engineered to insert a target gene fragment to be silenced. Both RNA1 and RNA2 cDNAs have been cloned into T-DNA cassette between duplicated 35S promoter and the nopaline synthase (NOS) terminator (13). Here we describe a detailed method of *Agrobacterium* infiltration-based VIGS assay in cotton seedlings. We further provide an example of using VIGS assay to understand gene functions in cotton seedling resistance to *Verticillium dahliae* infection. The protocol established here could be potentially adapted to study a diverse array of biotic and abiotic stress responses in cotton and provides a powerful tool in cotton functional genomics.

2. Materials

2.1. Plants, Growth Conditions, and Pathogen Strain

1. Cotton seeds: upland cotton (*Gossypium hirsutum*) variety FM9160 seeds obtained from Bayer CropScience (Lubbock, TX, USA).
2. Soil: Metro Mix 700 (SunGR, Beavile, WA, USA).
3. Growth room conditions: 23–25°C and 120 μE m^{-2} s^{-1} light, with a 12 h light/12 h dark photoperiod.
4. Pathogen strain: *Verticillium dahliae* (isolate King).

2.2. Plasmid Construction and Cloning

1. PCR amplification reagents: 10× reaction buffer, 10 mM dNTP, and Phusion high-fidelity DNA polymerase (New England BioLabs, MA, USA).
2. Restriction enzymes: EcoRI and KpnI (New England BioLabs, MA, USA).
3. DNA ligation kit: 10× T4 DNA ligase buffer and T4 DNA ligase (4 U/μl) (New England BioLabs, MA, USA).
4. VIGS RNA2 vector: pYL156 (pTRV2:RNA2).
5. QIAquick Gel Extraction Kit (QIAGEN).
6. LB liquid medium.
7. LB plates containing antibiotics.
8. Kanamycin (50 mg/ml stock) and gentamicin (50 mg/ml stock).
9. *Agrobacterium tumefaciens* GV3101 electro-competent cells stored in 10% glycerol at –80°C.

2.3. Agrobacterium Infiltration for VIGS Assay and Confirmation of Gene Silencing

1. *Agrobacterium tumefaciens* GV3101 containing pTRV1 (pTRV-RNA1).
2. *Agrobacterium* induction culture solution: LB liquid medium containing 50 μg/ml of kanamycin, 50 μg/ml of gentamicin, 10 mM of MES (2-(4 morpholino)-ethane sulfonic acid), and 20 μM acetosyringone.
3. *Agrobacterium* infiltration solution: 10 mM $MgCl_2$ containing 10 mM of MES and 200 μM acetosyringone.
4. 1 ml needleless syringes.
5. Syringe needles (20 Gauze).
6. Spectrum™ Plant Total RNA Kit (Sigma).
7. cDNA synthesis kit (Invitrogen).
8. PCR machine.

3. Methods

3.1. Grow Cotton Plants

1. Fill the soil in square pots (7 cm in diameter) and put the pots in a tray.
2. Sow the seeds in the soil (one seed per pot).
3. Soak the potting soil by pouring water in the tray.
4. Cover the tray with a plastic dome; grow the seedlings in the growth room until two cotyledons have emerged and remove the dome. Growth room condition: 23–25 °C and 120 μE m^{-2} s^{-1} light with a 12 h light/12 h dark photoperiod (see Note 1).

3.2. Clone Cla1 and Other Genes of Interest into pYL156 Vector

1. Search your genes of interest through blast against the *Gossypium* unigenes database at http://www.cottondb.org/blast/blast.html. Design a pair of primers that could amplify about 500 bp of target genes with EcoRI at 5′ end and KpnI at 3′ end. We used *Arabidopsis Cloroplastos alterados 1* (*AtCla1*, AT4G15560) as a query for the blast search for cotton *Cla1* gene (20) (see Notes 2 and 3).
2. A 500 bp fragment of *Cla1* gene or other genes of interest could be amplified by PCR with Phusion high-fidelity DNA polymerase from the cDNA library synthesized with RNA isolated from cotton leaf tissues (see Notes 4 and 5).
3. PCR products are purified with ethanol precipitation and digested with EcoRI and KpnI, together with pYL156 vector. The digestion is done in 15 μl mixture containing 0.2 μl of each enzyme, 1.5 μl of NEBuffer 4, and 0.15 μl of 100× BSA. The digestion is conducted at 37°C for 2–3 h.

4. Recover the digested vector and PCR fragments from DNA agarose gel using QIAquick Gel Recover Kit following the manufacturer's instruction.
5. Ligate the digested PCR products with vector: the ligation is done in a 15 μl mixture containing 1 μl of vector DNA (5–10 ng), 5 μl of insert DNA (20–50 ng), 1.5 μl of 10× T4 DNA ligase buffer, and 0.2 μl of T4 DNA ligase (4 U/μl). The reaction is incubated at room temperature for ~2 h.
6. Add 50 μl of *E. coli* competent cell into the ligation mixture, heat shock at 37°C for 2 min; chill on ice; add 200 μl LB liquid; and recover at 37°C for 30 min with a roller drum.
7. Spread the *E. coli* culture on LB plate containing 50 μg/ml kanamycin, and incubate the plate at 37°C overnight.
8. Screen the clones by miniprep DNA isolation and digestion of the DNA with EcoRI and KpnI. Finally, sequence the clones to confirm the insertion.
9. The confirmed clones are transformed into *Agrobacterium tumefaciens* GV3101 by electroporation method.
10. Spread the culture on LB plate containing kanamycin (50 μg/ml) and gentamicin (50 μg/ml); incubate at 28°C for 2 days. Culture the *Agrobacteria* with LB liquid, and store the culture with 25% glycerol at −80°C for further use.

3.3. VIGS Processes

1. Select a single colony from the fresh LB plates containing *Agrobacterium tumefaciens* carrying pTRV1, pYL156 (empty vector control), pYL156-GrCla1, and pYL156 carrying Your Favorite Gene (pYL156-YFG).
2. Inoculate the single colony with 5 ml of LB liquid containing kanamycin (50 μg/ml) and gentamicin (50 μg/ml); culture at 28°C overnight on a roller drum at 50 rpm.
3. Add 45 ml of *Agrobacterium* induction culture solution into the above culture; incubate at 28°C overnight in a shaker at 100 rpm.
4. Harvest the bacteria at 1,180 × *g* for 5 min, resuspend the pellet in *Agrobacterium* infiltration solution, and adjust the OD 600 to 1.5.
5. Leave the bacterial cultures on the bench at room temperature for 3 h (see Note 6).
6. Gently punch a couple of holes on the backside of the cotton cotyledons using a fine-tip needle without piercing through the tissue (see Note 7).
7. Mix the *Agrobacterial* culture suspension of pTRV1 with pYL156, pYL156-GrCla1, or pYL156-YFG at a 1:1 ratio.

8. Hand infiltrate the mixture into the cotyledons through the wounding sites using a needleless syringe.
9. Cover the infiltrated plants with a plastic dome and leave at room temperature overnight under the dim light (30 μE m^{-2} s^{-1}) (see Note 8).
10. Transfer the plants to the growth room and remove the dome.
11. About 7–10 days later, examine the silencing phenotype for *Cla1* or *YFG*. The true leaves on the plants infiltrated with pYL156-GrCla1 will show albino phenotype (see Note 9).
12. Harvest the true leaves from silenced and control plants; isolate RNA with Spectrum™ Plant Total RNA Kit (Sigma).
13. Synthesize cDNA with cDNA synthesis kit (Invitrogen), and perform semiquantitative or real-time RT-PCR analysis to confirm the silencing of endogenous genes (see Note 10).

3.4. Verticillium Infection of VIGS Silenced Cotton Plants

1. Culture *Verticillium dahliae* (isolate King) on potato dextrose agar (PDA) plates at room temperature (23°C) for 3–4 days.
2. When pYL156-*GrCla1*-silenced plants show albino symptoms, prepare the spore suspension of *V. dahliae.*
3. Collect the fungal spores by scratching the fungal mycelium on the plate surface using sterile dH_2O. The spores are filtered with autoclaved cheesecloth, counted under a microscope using a hemacytometer, and diluted to the concentration of 1×10^6 /ml in sterile H_2O containing 0.001% Tween 20.
4. Infiltrate 100 μl of spore suspension to the stems of silenced cotton plants at a position approximately 1 cm below cotyledons using a 22 Gauze needle syringe (see Note 11).
5. Score the Verticillium wilting phenotype by examining the wilting symptoms appeared on the true leaves of plants (see Notes 12–14).

4. Notes

1. Maintaining the temperature of growth room at approximately 23–25°C and light at 120 μE m^{-2} s^{-1} is important to obtain a consistent and uniform silencing efficiency. The cotton seedlings are also easy to be infiltrated with *Agrobacteria* when grown under these conditions (14, 20).
2. More than one gene of interest could be inserted into pYL156 vector to silence multiple genes at once in one vector. The DNA fragment for each gene is about 200–300 bp to be inserted into the vector. A three-way ligation could be used to insert two DNA fragments in the vector. According to our

experience with *Arabidopsis*, efficiency of silencing two genes is higher with *Agrobacteria* containing two genes inserted into one vector than the mixture of *Agrobacteria* with individual genes from two vectors.

3. The primer for cloning cotton *Cla1* gene into pYL156 vector is *Cla1-F*, 5′-GGAATTCCACAACATCGATGATTTAG-3′, *Cla1-R*, 5′-GGGGTACCATGATGAGTAGATTGCAC-3′. The PCR product should not contain internal EcoRI and KpnI sites.
4. We cloned cotton *Cla1* from *G. raimondii* cDNA library, and it worked very well in all upland *G. hirsutum* cottons we tested, suggesting the high sequence conservation among different cotton genomes.
5. The DNA fragment used for silencing ranges from 250 to 500 bp in length. Further shortening the length of fragment will reduce the silencing efficiency.
6. Incubation for 3 h at room temperature will enhance the silencing efficiency. It can be extended up to 24 h.
7. Direct infiltration of the *Agrobacterial* culture into the cotton cotyledons often results in severe wounding effect. Punching small holes on the backside of cotyledons greatly facilitates the infiltration and reduces the wounding damage.
8. After infiltration, covering the plants with a dome to keep the high humidity under dim light and relatively low temperature (21–23°C) facilitates the plants to recover from the wounding effects and the *Agrobacterial* infiltration.
9. The albino phenotype on the newly emerging leaves will be observed approximately 10 days post-infiltration. One month later, 100% of *GrCla1-Agrobacteria*-infiltrated plants will exhibit a strong albino phenotype. This was repeatedly observed in many varieties of upland cotton (20).
10. The primers for cotton *Cla1* semiquantitative RT-PCR are *Cla1-F*, 5′-GCCCTTTGTGCATCTTC-3′, *Cla1-R*, 5′-CTC TAGGGGCATTGAAG-3′. *GhActin9* could be used as an internal standard control (20).
11. There are several other approaches available for *Verticillium* infection, such as uprooted dip inoculation (21) and cotyledonary node drop inoculation (22). When performing the stem inoculation, it is important to pierce through halfway of the stem and infiltrate the spore suspension into the stem.
12. It usually takes 10–14 days to observe the *Verticillium* wilting phenotype after inoculation. However, the symptom development depends on the varieties, plant growth conditions, and pathogenicity of *Verticillium*. The wilting phenotype can be examined by either counting the percentage of wilting plants or measuring the stunting appearance (20).

13. By using VIGS approach, we have demonstrated that *GhNDR1* and *GhMKK2* are required for *Verticillium* resistance in cotton (20).
14. The *Agrobacterium*-infiltrated plants should be kept under contained quarantine conditions, and plants should be properly autoclaved and disposed after the assays as TRV has a broad host range and is a notifiable pathogen.

Acknowledgments

We thank Dr. S. P. Dinesh-Kumar for pTRV-VIGS vectors and Dr. Terry Wheeler and Bayer CropScience (Lubbock, TX, USA) for cotton seeds. This work was supported by Texas AgriLife Research Cotton Improvement Program to L. S.

References

1. Sunilkumar G, Campbell LM, Puckhaber L et al (2006) Engineering cottonseed for use in human nutrition by tissue-specific reduction of toxic gossypol. Proc Natl Acad Sci USA 103:18054–18059
2. Yu J, Kohel RJ, Smith CW (2010) The construction of a tetraploid cotton genome wide comprehensive reference map. Genomics 95:230–240
3. Zhang HB, Li Y, Wang B, Chee PW (2008) Recent advances in cotton genomics. Int J Plant Genomics 2008:742304
4. http://www.fas.usda.gov/wap/circular/2010/10-05/productionfull05-10.pdf
5. Fradin EF, Thomma BP (2006) Physiology and molecular aspects of Verticillium wilt diseases caused by V. dahliae and V. alboatrum. Mol Plant Pathol 7:71–86
6. Bolek Y, El-Zik KM, Pepper AE et al (2005) Mapping of verticillium wilt resistance genes in cotton. Plant Sci 168:1581–1590
7. Yang C, Guo WZ, Li GY, Gao F, Lin SS, Zhang TZ (2008) QTLs mapping for Verticillium wilt resistance at seedling and maturity stages in *Gossypium barbadense* L. Plant Sci 174: 290–298
8. Hashmi JA, Zafar Y, Arshad M et al (2011) Engineering cotton (*Gossypium hirsutum* L.) for resistance to cotton leaf curl disease using viral truncated AC1 DNA sequences. Virus Genes 42:286–296
9. Chen ZJ, Scheffler BE, Dennis E et al (2007) Toward sequencing cotton (Gossypium) genomes. Plant Physiol 145:1303–1310
10. Burch-Smith TM, Anderson JC, Martin GB et al (2004) Applications and advantages of virus-induced gene silencing for gene function studies in plants. Plant J 39:734–746
11. Dinesh-Kumar SP, Anandalakshmi R, Marathe R et al (2003) Virus-induced gene silencing. Methods Mol Biol 236:287–294
12. Becker A, Lange M (2010) VIGS–genomics goes functional. Trends Plant Sci 15:1–4
13. Hayward A, Padmanabhan M, Dinesh-Kumar SP (2011) Virus-induced gene silencing in *Nicotiana benthamiana* and other plant species. *Plant Reverse Genetics: Methods and Protocols.* Methods Mol Biol 678:55–63
14. Tuttle JR, Idris AM, Brown JK et al (2008) Geminivirus mediated gene silencing from Cotton leaf crumple virus is enhanced by low temperature in cotton. Plant Physiol 148: 41–50
15. Kaloshian I (2007) Virus-Induced Gene silencing in plants roots. *Plant-Pathogen Interactions*: Methods and Protocols. Methods Mol Biol 354:173–181
16. Pflieger S, Blanchet S, Camborde L et al (2008) Efficient virus-induced gene silencing in *Arabidopsis* using a 'one-step' TYMV-derived vector. Plant J 56:678–690
17. Ding XS, Schneider WL, Chaluvadi SR et al (2006) Characterization of a Brome mosaic virus strain and its use as a vector for gene silencing in monocotyledonous hosts. Mol Plant Microbe Interact 19:1229–1239
18. Scofield SR, Huang L, Brandt AS et al (2005) Development of a virus induced gene-silencing

system for hexaploid wheat and its use in functional analysis of the Lr21-mediated leaf rust resistance pathway. Plant Physiol 138: 2165–2173

19. Mei CS, Zhou X, Yang Y (2007) Use of RNA interference to dissect defense-signaling pathways in rice. *Plant-Pathogen Interactions*: Methods and Protocols. Methods Mol Biol 354:161–172

20. Gao X, Wheeler T, Li Z et al (2011) Silencing GhNDR1 and GhMKK2 compromised cotton resistance to *Verticillium wilt*. Plant J 66(2): 293–305

21. Ellendorff U, Fradin EF, de Jonge R et al (2009) RNA silencing is required for Arabidopsis defence against Verticillium wilt disease. J Exp Bot 60:591–602

22. Parkhi V, Kumar V, Campbell LM et al (2010) Resistance against various fungal pathogens and reniform nematode in transgenic cotton plants expressing Arabidopsis *NPR1*. Transgenic Res 19:959–975

Chapter 13

Highly Efficient Virus-Induced Gene Silencing in Apple and Soybean by *Apple Latent Spherical Virus Vector* and Biolistic Inoculation

Noriko Yamagishi and Nobuyuki Yoshikawa

Abstract

Virus-induced gene silencing (VIGS) is an effective tool for the analysis of the gene function in plants within a short time. However, in woody fruit tree like apple, some of Solanum crops, and soybean, it is generally difficult to inoculate virus vector by conventional inoculation methods. Here, we show efficient VIGS in apple and soybean by *Apple latent spherical virus* (ALSV) vector and biolistic inoculation. The plants inoculated with ALSV vectors by particle bombardment showed uniform silenced phenotypes of target genes within 2–3 weeks post inoculation.

Key words: Apple latent spherical virus (ALSV), Virus-induced gene silencing (VIGS), Fruit tree, Apple, Soybean, Biolistic inoculation, Efficient inoculation method

1. Introduction

Virus-induced gene silencing (VIGS) is an attractive tool for functional plant genomics, and various kinds of VIGS vectors have been reported and applied to research of functional genomics in plants around the world (1, 2). The reliability and effectiveness of VIGS depends on both plant species and characteristics of virus vectors (3–7).

Apple latent spherical virus (ALSV), originally isolated from an apple tree, has isometric virus particles ca. 25 nm in diameter, and it contains two ssRNA species (RNA1: 6813nt and RNA2: 3385 nt) and three capsid proteins (Vp25, Vp20, and Vp24) (8, 9). The characteristics of ALSV are suitable for the virus vector because the virus has a wide host range, does not induce any obvious symptoms, and infects uniformly and systemically in most host plants.

Annette Becker (ed.), *Virus-Induced Gene Silencing: Methods and Protocols*, Methods in Molecular Biology, vol. 975, DOI 10.1007/978-1-62703-278-0_13, © Springer Science+Business Media New York 2013

ALSV vectors have been constructed for the expression of foreign genes in plants (9, 10), and have been shown to be used for a reliable and effective VIGS among a broad range of plants, including legumes and *Rosaceae* fruit trees (11–14).

To establish the VIGS system in high-throughput functional genomics in plants, stable and efficient inoculation method of viral vector to plants is a matter of great importance. In most herbaceous host species of ALSV like *Arabidopsis thaliana*, *Nicotiana* spp., *Petunia hybrida*, *Cucurbitaceae* plants, and *Chenopodium quinoa* (*C. quinoa*), it is easy to inoculate ALSV vectors by rub inoculation with sap of ALSV vector-infected leaves (9–11, 14). On the other hand, in woody fruit trees like apple, some *Solanum* crops, and soybean, it is generally difficult to achieve a highly efficient infection rate of ALSV vectors by sap inoculation. Here, we show VIGS using ALSV vector and biolistic inoculation using a Helios Gen Gun System (Bio-Rad). Infected apple and soybean show uniform silencing phenotypes.

2. Materials

All solutions are prepared using RNase-free ultrapure water (prepared by purifying deionized water to attain a sensitivity of 18 MΩ cm at 25°C) and molecular biological grade reagents.

2.1. Plants

1. Seeds of *C. quinoa* are planted in pots and grown in a greenhouse (18–25°C, long-day condition) until seven true-leaf stages.
2. Seeds of apple (*Malus x domestica*) are stored in a dampened paper towel at 4°C until germination.
3. Seeds of soybean (*Glycine max*) are planted in a petri dish filled with damped vermiculite at 25°C until germination.

2.2. Construction of Infectious cDNA Clone Containing Target Sequence

1. Infectious cDNA clone of ALSV-RNA2 (pEALSR2L5R5).
2. A cDNA clone of *ribulose-1,5-bisphosphate carboxylase small subunit* (*rbcS*) of apple and a cDNA clone of *phytoene desaturase* of soybean (*soyPDS*).
3. Two primer pairs for amplification of the 201 bp of *rbcS* of apple (rbcSXho(+): 5′-TACATCTCGAG^{250}AGGAAGGTAAGA GAGGGT^{233}-3′, containing *Xho* I site (underlined) and rbcS-201Bam(−): **5′-TACATGGATCC**^{250}AGGAAGGTAAGA GAGGGT^{233}-3′, containing *Bam*HI site (underlined)) and the 300 bp of *soyPDS* (soyPDS494Xho(+): **5′-TACATCTCGAG**^{494} TCTCCGCGTCCTCTAAAA^{511}-3′, containing *Xho*I site (underlined) and soyPDS793Bam(−): **5′-TACATGGATCC**^{793} TCCAGGCTTATTTTGGCAT^{776}-3′, containing *Bam*HI site (underlined)).

4. *Xho*I and *Bam*HI and suitable buffer for each enzyme.
5. Taq DNA polymerase, Taq DNA polymerase buffer, dNTPs for Polymerase chain reaction (PCR).
6. DNA Ligation Kit (Takara, Kyoto, Japan).
7. Competent cells for DNA cloning: *Escherichia coli* DH5α or JM109.
8. SOC medium: Shake until the solutes (20 g bacto-tryptone, 5 g bacto-yeast extract, 0.58 g NaCl, and 0.19 g KCl) have dissolved. Adjust the volume to 980 mL with deionized water and sterilize by autoclave. After the medium has been autoclaved, allow it to cool down, and then add 10 mL of sterile solution of 1 M $MgCl_2 \cdot 6H_2O$, 10 mL of sterile solution of 1 M $MgSO_4 \cdot 6H_2O$, and 10 mL of sterile solution of 2 M glucose. Store at –20°C.
9. LB agar plate: Shake until the solutes (1 g bacto-tryptone, 0.5 g bacto-yeast extract, and 1 g NaCl) have dissolved and adjust the pH 7.5 with 3N NaOH. Adjust final volume to 100 mL with deionized water. Add 1.5 g agar, sterilize by autoclave. Allow the LB medium to cool to about 50°C, and add ampicillin (50 mg/mL). Mix the LB medium by swirling and pour in the plates. LB agar plates are stored at 4°C.
10. Ampicillin (50 mg/mL): ampicillin is made by dissolving in deionized water and sterilized by filtration through a 0.22-μm filter. Store at –20°C.

2.3. Purification of cDNA Clone

1. LB medium: Shake until the solutes (10 g bacto-tryptone, 5 g bacto-yeast extract, and 10 g NaCl) have dissolved and adjust the pH 7.5 with 3N NaOH. Adjust final volume to 1,000 mL with deionized water. Sterilize by autoclave. Store at 4°C.
2. QIAGEN Plasmid Midi Kit (QIAGEN, Duesseldorf, Germany).
3. 2-Propanol.
4. 70% Ethanol.
5. TE buffer: After the 99.3 mL of deionized water has been autoclaved, allow it to cool down, and add 0.5 mL of sterile solution of 2 M Tris–HCl (pH 8.0) and 0.2 mL of sterile solution of 0.5 M EDTA (pH 8.0) to 99.3 mL of sterile and deionized water. Store at room temperature.

2.4. Inoculation to C. quinoa

1. Purified infectious cDNA clones (pEALSR1 and pEALSR2L5R5).
2. *C. quinoa* (6–8 leaf stages).
3. 600-mesh carborundum (NAKARAI TESQU INC., Kyoto, Japan).
4. Latex finger cots.

2.5. Inoculation to Soybean and Apple

1. Infected *C. quinoa* leaves.
2. Mortar and pestle.
3. Tri reagent (Sigma-aldrich Japan k.k., Tokyo, Japan) or Tripure isolation reagent (Roche Applied Science, Basel, Switzerland).
4. H_2O-desaturated phenol–chloroform (1:1).
5. 99.5% ethanol.
6. 80% ethanol.
7. RNase-free H_2O.
8. 5 M ammonium acetate.
9. 2-propanol.
10. 100% ethanol dehydrated with molecular sieves.
11. Helios Gene Gun system (Bio-Rad, Hercules, CA, USA).

2.6. RNA Analysis of ALSV Vector Infected Apple

1. RNA extraction buffer: 2% (w/v) cetyltrimethylammonium bromide (CTAB), 2% polyvinylpolypyrrolidone (PVP), 100 mM Tris–HCl (pH 8.0), 25 mM (pH 8.0), 2 M NaCl, 2% β-mercaptoethanol.
2. Chloroform.
3. 7.5 M LiCl: LiCl (31.8 g) is dissolved in deionized water, adjusting final volume to 100 mL, and sterilized by autoclave. Store at room temperature.
4. 80% Ethanol.
5. RNase-free water.
6. ReverTra Ace reverse transcriptase (TOYOBO, Osaka, Japan).
7. oligo (dT) primer.
8. Taq DNA Polymerase with ThermoPol Buffer (New England Biolab (NEB) Ltd., Hitchin, UK).
9. A primer pair for detection of ALSV: R2ALS+primer (5′-^{1362}GCGAGGCACTCCTTA^{1376}-3′) and R2ALS-(5′-^{1524}GCAAGGTGGTCGTGA^{1510}-3′), which were designed for the amplification of a specific region containing the insert sequence on the ALSV RNA2 genome.

2.7. RNA Analysis of ALSV Vector Infected Soybean

1. Nylon membrane for RNA fixation; Hybond-N+(GE Healthcare bioscience, NJ, USA).
2. 0.05 N NaOH: NaOH (1 g) is dissolved in deionized water, adjusting final volume to 500 mL. Make just before use.
3. 20× SSC: Dissolve the solutes (175.3 g NaCl and 88.2 g Sodium citrate) and adjust the pH 7.0. Adjust final volume to 1,000 mL with deionized water. Sterilize by autoclave. Store at room temperature.
4. Formamide.

5. 10% Blocking reagent (Roche Applied Science, Basel, Switzerland): Blocking reagent (5 g) is dissolved in buffer 1 (100 mM Maleic acid and 150 mM NaCl; pH 7.5), adjusted to a final volume of 50 mL, and sterilized by autoclave. Store at –20°C.
6. 10% Sarcosyl: Sarcosyl (5 g) is dissolved in deionized water, adjusted to a final volume of 50 mL and sterilized by filtration through a 0.45-μm filter. Store at room temperature.
7. 10% SDS: SDS is dissolved in deionized water, adjusted to a final volume of 50 mL and sterilized by filtration through a 0.45-μm filter. Store at room temperature.
8. DIG-labeled RNA probes for detection of ALSV vector: DIG-labeled RNA probes complementary to positions 1–476 of ALSV-RNA1 (Genbank/EMBL/DDBJ accession no. AB030940) and 1–433 of ALSV-RNA2 (Genbank/EMBL/DDBJ accession no. AB030941) (Fig. 1).
9. Ambion MEGAscript T7 kit (Ambion, Applied Biosystems, Foster City, CA, USA).
10. DIG RNA labeling mix (Roche Applied Science, Basel, Switzerland).
11. Anti-digoxigenin-AP, Fab fragments (Roche Applied Science, Basel, Switzerland).
12. CDP-*Star* (Roche Applied Science, Basel, Switzerland).
13. X-ray films.

3. Methods

Carry out all procedures at room temperature unless otherwise specified.

3.1. Plants for Propagation of ALSV Vector

Seeds of *C. quinoa* are planted in pots and grown in a greenhouse (long-day condition; 18–25°C) until 6–8 true-leaf stage.

3.2. Construction of Apple Latent Spherical Virus Vectors Inducing VIGS of soyPDS and rbcS of Apple

1. ALSV vectors inducing VIGS of *soyPDS* and *rbcS* of apple are constructed by using ALSV-RNA2 infectious clone (pEALSR2L5R5) (Fig. 1) (see Note 1).
2. *soyPDS* and *rbcS* DNA fragment is amplified from a cloned cDNA of *soyPDS* or *rbcS* gene by using the primer pair described above.
3. The DNA product is double-digested with *Xho*I and *Bam*HI and ligated to pEALSR2L5R5 restricted with the same enzymes.
4. Competent *E. coli* DH5α or JM109 cells are transformed by the plasmids.

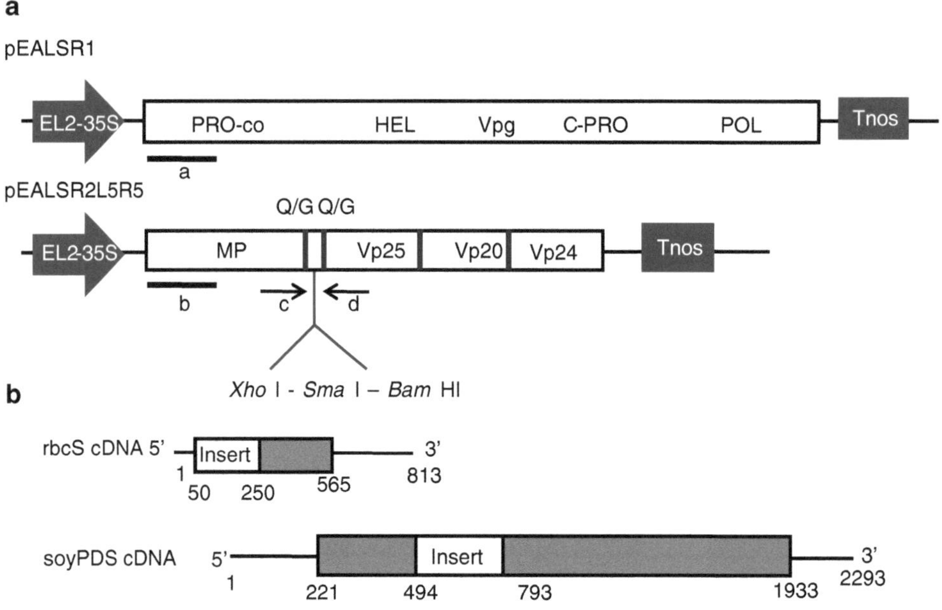

Fig. 1. (**a**) Schematic representation of the infectious cDNA clones of RNA1 (pEALSR1) and RNA2-based vector (pEALSR2L5R5). pEALSR2L5R5 was constructed by creating an artificial protease processing site by duplicating the Q/G protease cleavage site between MP and Vp25. Cloning sites *Xho*I, *Sma*I, and *Bam*HI for foreign genes between duplicating the Q/G protease cleavage sites are shown. (**b**) Schematic representation of full-length *rbcS*- and *soyPDS* cDNAs of apple and soybean. Gray boxes and with nucleotide numbers indicate the coding regions of each gene. The *rbcS*- and *soyPDS* cDNA regions inserted into pEALSR2L5R5 are represented as *white boxes* with bold nucleotide numbers. *Bold black lines* (a and b) indicate the regions of probes used for Tissue-blot analysis to detect ALSV-RNAs. *Arrows* (c and d) indicate the positions of a primer pair for detection of ALSV RNAs by RT-PCR. P35S, enhanced CaMV 35S promoter; Tnos, nopaline synthase terminator; PRO-co, protease cofactor; HEL, NTP-binding helicase; C-PRO, cysteine protease; POL, RNA polymerase; MP, 42K movement protein; Vp25, Vp20, and Vp24, capsid proteins.

5. Add 1 mL SOC medium to transformed competent *E. coli* tubes and incubate the tubes in a shaking incubator at 37°C for 1 h.
6. The transformed competent cells are concentrated by centrifugation at room temperature and transferred onto LB agar plates uniformly.
7. Incubate the plates at 37°C overnight.
8. The *E. coli* colonies are selected by antibiotic sensitivity and cultured in 2 mL of LB medium at 37°C overnight.
9. The plasmid DNAs are isolated from the cultured cells by the alkaline lysis procedure.
10. To identify the inserts, plasmid DNAs are restricted with *Xho*I and *Bam*HI and electrophoresed on a 1.5% agarose gel.
11. The resulting pEALSR2L5R5-based vectors containing 201 bp of *rbcS* or 300 bp of *soyPDS* genes are designated pEALSR2L-5R5rbcS and pEALSR2L5R5soyPDS, respectively.

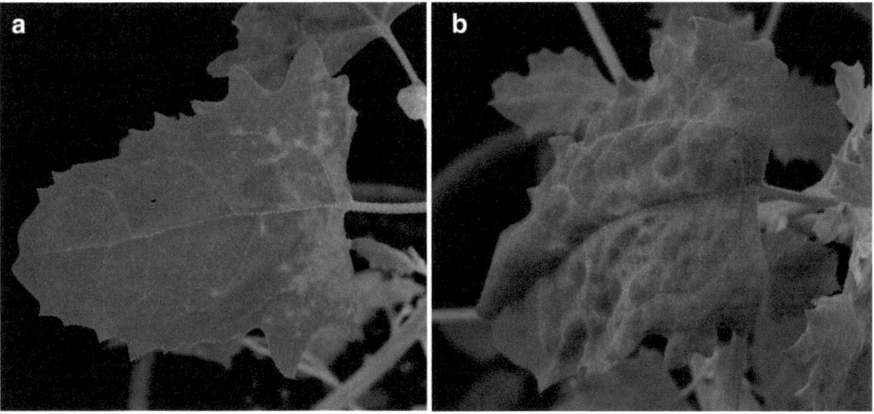

Fig. 2. Symptoms of *C. quinoa* infected with ALSV cDNA infectious clones. The symptoms consisting of chlorotic spots (**a**) and mosaic (**b**) appeared in the upper uninoculated leaves at about 7–14 days post inoculation.

3.3. Purification of Infectious cDNA Clones

1. Plasmids, pEALSR1 which is an infectious clone of ALSV-RNA1, pEALSR2L5R5rbcS, and pEALSR2L5R5soyPDS, are propagated in *E. coli* DH5α or JM109 cells.
2. A single colony of transformed *E. coli* is inoculated in a 50 mL of LB medium containing 50 μg/mL of ampicillin and cultured at 37°C for overnight with vigorous shaking.
3. Plasmid DNAs are purified by the QIAGEN Plasmid Midi Kit according to the manufacture's protocol, dissolved in TE buffer at a concentration of 1 μg/μL and stored at -20°C until use.

3.4. Inoculation to Plants

3.4.1. Inoculation of Infectious cDNA Clones to C. quinoa

1. Both purified pEALSR1 and pEALSR2L5R5rbcS (or pEALSR-2L5R5soyPDS) are mixed at a concentration of 1 μg/μL each.
2. *C. quinoa* leaves are dusted with 600-mesh carborundum (NAKARAI TESQU INC. Kyoto, Japan).
3. The mixture of pEALSR1 and pEALSR2L5R5rbcS (or pEALSR2L5R5soyPDS) (8 μL) is dropped onto the tip of a finger wrapped with latex finger cots and mechanically inoculated onto a leaf of *C. quinoa* plant (seven true-leaf stage). The inoculation of infectious cDNA clones is conducted on four leaves (third to sixth true leaves) per plant.
4. A surface of inoculated leaves is flushed with a sufficient amount of water to remove carborundum.
5. The symptoms consisting of chlorotic spots and mosaic appeared in the upper uninoculated leaves 7–10 days post inoculation (dpi) (Fig. 2). The resulting ALSV vectors derived from pEALSR2L5R5rbcS and pEALSR2L5R5soyPDS are designated as rbcS201-ALSV and soyPDS-ALSV, respectively.
6. *C. quinoa* leaves with symptoms are collected and stored at -80°C (see Note 2).

3.4.2. Biolistic Inoculation of rbcS201-ALSV onto Apple Cotyledons of Seeds Just After Germination

1. Total RNAs are extracted from rbcS201-ALSV infected *C. quinoa* leaves by Tri reagent (Sigma-aldrich Japan k.k., Tokyo, Japan) or Tripure isolation reagent (Roche Applied Science, Basel, Switzerland) according to the instruction manuals. Total RNAs are re-extracted by phenol–chloroform and then precipitated with ethanol, washed by 80% ethanol, and dissolved in RNase-free H_2O at a concentration of 4 μg/μL.
2. Weigh out 8 mg of gold particles (0.6 μm in diameter) into a 1.5 mL microcentrifuge tube.
3. Add 50 μL RNase-free H_2O to a tube.
4. Vortex a tube vigorously with a MicroMixer E-36 (Taitec, Saitama, Japan) for 5 min and then sonicate a tube for at least 5 min.
5. While vortexing a tube with a MicroMixer E-36, 50 μL of total RNAs solution extracted from rbcS201-ALSV infected *C. quinoa* leaves, 10 μL of 5 M ammonium acetate, and 220 μL of 2-propanol are added sequentially.
6. After continual vortexing for 5–10 min, gold particles coated with the RNAs are placed inside at −20°C for at least 1 h.
7. The pellet of gold particles is gently washed three times with 1 mL of 100% ethanol.
8. Resuspend the pellet in 0.2 mL of 100% ethanol by tapping the side of the tube (see Note 3) and move the suspension to a 15 mL tube.
9. Add 2.2 mL of 100% ethanol and resuspend well the gold particles in ethanol.
10. The suspension is used for the preparation of the gold-coated tubing for biolistic inoculation according to the manufacture's instructions.
11. The tube coated with RNAs is cut into 40 cartridges, each containing 0.2 mg gold particles coated with about 5.0 μg total RNAs extracted from rbcS201-ALSV infected *C. quinoa* leaves.
12. Biolistic inoculation of ALSV onto apple is performed to cotyledons of seeds just after germination. The seed coat is removed just before inoculation (Fig. 3a) (see Note 4).
13. The cotyledons of apple seeds are covered with a wire-netting to fix them due to the pressure of helium gas and are bombarded with gold particles coated with total RNAs from rbcS201-ALSV infected leaves at a pressure of about 300 psi using a Helios Gene Gun system (Bio-Rad). Cotyledons of the plants are bombarded with two shots per cotyledon.
14. After particle bombardment, germinated seeds are sprayed with water and placed on a KIMWIPE (Nippon Paper Crecia, Tokyo, Japan) soaked with water to retain humidity in the petri dish. The petri dish is placed at 4°C in the dark for 2 days.

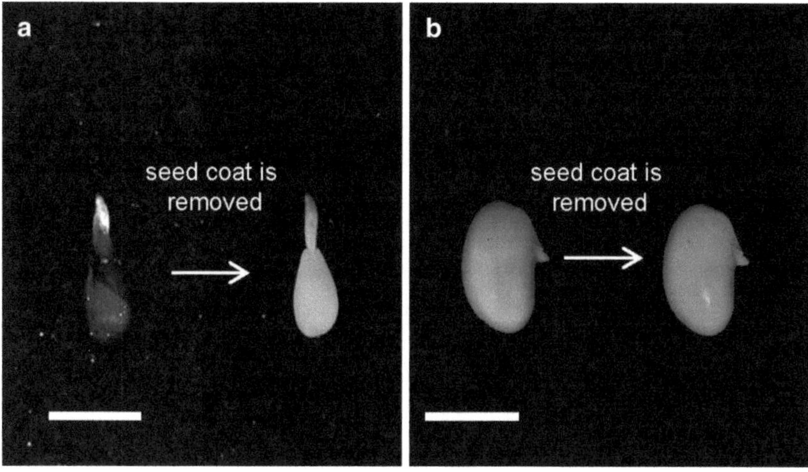

Fig. 3. Apple and soybean for biolistic inoculation. Seed of apple (**a**; *left*) and soybean (**b**; *left*) used for biolistic inoculation. The seed coats are removed just before biolistic inoculation (**a** and **b**; *right*). Scale bar: 1 cm.

15. After 2 days, The petri dish is wrapped with paper to lower light intensity and placed into a growth chamber (25°C, 16:8 light–dark conditions) for 1 day. Next day, the paper is removed, and the apple seedlings are grown under the same conditions for 3 days.
16. After 3 days, the seedlings are planted in soil, acclimatized to the open air, and grown under the same condition.

3.4.3. Inoculation of soyPDS-ALSV onto Soybean Cotyledons of Germinated Seeds

1. Total RNAs are extracted from soyPDS-ALSV infected *C. quinoa* leaves by Tri reagent (Sigma-aldrich Japan k.k., Tokyo, Japan) or Tripure isolation reagent (Roche Applied Science, Basel, Switzerland) according to the instruction manuals. They are re-extracted by phenol–chloroform and then precipitated with ethanol, washed by 80% ethanol, and dissolved in RNase-free H_2O at a concentration of 4 μg/μL.
2. Weigh out 8 mg of gold particles (0.6 μm in diameter) into a 1.5 mL microcentrifuge tube.
3. Add 50 μL RNase-free H_2O to the tube.
4. Vortex the tube vigorously with a MicroMixer E-36 (Taitec, Saitama, Japan) for 5 min and then sonicate the tube for at least 5 min.
5. While vortexing the tube with a MicroMixer E-36, 50 μL of total RNAs solution extracted from rbcS201-ALSV infected *C. quinoa* leaves, 10 μL of 5 M ammonium acetate, and 220 μL of 2-propanol are added sequentially.
6. After continual vortexing for 5–10 min, gold particles coated with the RNAs are placed inside at −20°C for at least 1 h.
7. The pellet of gold particles is gently washed three times with 1 mL of 100% ethanol.

8. Resuspend the pellet in 0.2 mL of 100% ethanol by tapping the side of the tube (see Note 4) and move the suspension to a 15 mL tube.
9. Add 2.2 mL of 100% ethanol and resuspend well the gold particles in ethanol.
10. The suspension is used for the preparation of the gold-coated tubing for biolistic inoculation according to the manufacture's instructions.
11. The tube coated with RNAs is cut into 40 cartridges, each containing 0.2 mg gold particles coated with about 5.0 μg total RNAs extracted from soyPDS-ALSV infected *C. quinoa* leaves.
12. Biolistic inoculation of ALSV is performed to cotyledons of seeds at an emergence stage. The seed coat is removed just before inoculation (Fig. 3b).
13. The cotyledons of soybean seed at an emergence stage is covered with wire-netting to fix them due to the pressure of helium gas and bombarded with gold particles coated with total RNAs from soyPDS-ALSV infected leaves at a pressure of about 300 psi using a Helios Gene Gun system (see Note 4). Cotyledons of the plants are bombarded with 2–3 shots per cotyledon.
14. After particle bombardment, germinated seeds are sprayed with water and placed on a KIMWIPE (Nippon Paper Crecia, Tokyo, Japan) soaked with water to retain humidity in the petri dish. The petri dish is placed into a growth chamber (25°C, 16:8 light–dark conditions) for 3 days.
15. After 3 days, the seedlings are planted in soil, acclimatized to the open air, and grown under the same condition.

3.5. RNA Analysis of ALSV-Vector Infected Plants

3.5.1. rbcS201-ALSV Detection from Apple Leaves by RT-PCR

1. Fifty mg of rbcS201-inoculated apple leaves are homogenized with 500 μL extraction buffer (2% (w/v) cetyltrimethylammonium (CTAB), 2% polyvinylpolypyrrolidone (PVP), 100 mM Tris–HCl (pH 8.0), 25 mM EDTA (pH 8.0), 2 M NaCl, 2% β-mercaptoethanol) in a Micro Smash MS-100 bead beater (TOMY, Tokyo, Japan).
2. The homogenates are incubated at 65°C for 15 min and then mixed with 500 μL chloroform vigorously for 2 min.
3. After centrifugation at 6,708 × *g* for 10 min, the aqueous phases are moved into new 1.5 mL tubes.
4. Add one-third volume of 7.5 M LiCl to the aqueous phases and mixed well.
5. Incubate at –80°C for 30 min or at 4°C overnight.
6. After centrifugation at 13,147 × *g* for 30 min, RNA pellets were washed with 80% ethanol.

7. RNA pellets are dissolved in RNase-free water at a concentration of 1 μg/μL.
8. First strand cDNA is synthesized using 1 μg RNA, oligo(dT) primer, and ReverTra Ace reverse transcriptase.
9. PCR amplification is performed using 1 μL template cDNAs, a primer pair (R2ALS+ and R2ALS–), and Thermopol Taq polymerase.
10. The PCR reaction conditions are as follows: initial denaturation at 95°C for 1 min followed by 30 cycles of denaturation at 94°C for 20 s, annealing at 58°C for 10 s, and extension at 72°C for 30 s.
11. The products are separated on a 1% agarose gel and visualized by staining with ethidium bromide.

3.5.2. SoyPDS-ALSV Detection from Soybean Leaves by Tissue-Blot Hybridization

1. Soybean leaves infected with soyPDS-ALSV are freeze-thawed and then printed onto a Hybond-N+ membrane using a roller.
2. The leaves-printed membrane were treated with 0.05 N NaOH (30 min), 20× SSC (30 min), and baked in an oven (80°C, 2 h) for RNA fixation to the membrane.
3. DIG-labeled RNA probes for ALSV detection are synthesized by using Ambion MEGAscript T7 kit according the instruction manual except for replacement of NTPs of Megascript T7 kit with DIG RNA labeling MIX.
4. Prehybridization (2 h) and hybridization (18 h) of the membrane were carried out at 68°C in a hybridization solution (50% formamide, 5× SSC, 2% blocking reagent, 0.1% sarcosyl, and 0.02% SDS).
5. The membrane is washed twice for 5 min with 2× SSC, 0.1% SDS, and twice for 15 min with 0.1× SSC, 0.1% SDS at 68°C.
6. Chemiluminescent detection was conducted by anti-digoxigenin-AP, Fab fragments and CDP-*Star* Chemiluminescent substrate according the manufacturer's protocol.
7. The membrane was then exposed to X-ray films for 30 min.

3.6. VIGS in ALSV-Vector Infected Plants

3.6.1. VIGS in Apple Seedlings Infected with rbcS201-ALSV

1. The biolistic inoculation with total RNAs extracted from rbcS201-ALSV infected *C. quinoa* leaves to apple cotyledons results in high infection efficiency. In most case, more than 90% of inoculated apple seedlings can be infected (14–16). RbcS201-ALSV can be detected from non-inoculated upper leaves of apple seedlings by RT-PCR analysis (Fig. 4a).
2. The infected apple seedlings start to develop chlorosis on the 1st to 3rd true leaves from about 2 weeks post inoculation wpi (Fig. 4b). Newly developed true leaves show a more highly uniform chlorosis (Fig. 4c). Non-infected or wild-type ALSV (wtALSV) infected seedlings do not show any viral symptoms nor chlorosis (Fig. 4b and c).

Fig. 4. VIGS of rbcS gene in apple by rbcS201-ALSV infection. (**a**) Detection of rbcS201-ALSV by RT-PCR analysis. *Lane 1*, DNA marker (λ DNA/*Hind* III digest); *Lane 2*, non-inoculated apple leaf; *Lane 3*, rbcS-ALSV-inoculated apple leaf; *Lane 4*, H_2O as a negative control; *Lane 5*, pEALSR2L5R5rbc201 as a positive control. (**b**) Normal seedling infected with wtALSV (*left*) and leaf chlorosis in an apple seedling infected with rbcS201-ALSV (*right*) at 2nd true leaf stage. (**c**) Normal seedling infected with wtALSV (*left*) and leaf chlorosis in an apple seedling infected with rbcS201-ALSV (*right*) 33 dpi. Newly developed leaves of infected apple seedling show uniform chlorosis.

3.6.2. VIGS in Soybean Seedlings Infected with soyPDS-ALSV

1. The biolistic inoculation with total RNAs extracted from soyPDS-ALSV infected *C. quinoa* leaves to soybean cotyledons results in high infection efficiency. In most case, almost 100% of inoculated soybean seedlings can be infected (13).
2. The infected soybean seedlings show mosaic symptom on unifoliate and first trifoliate leaves. However, mosaic symptom are no longer observed above the 2nd or 3rd trifoliate leaves and start to develop photo-bleaching on unifoliate and first trifoliate leaves at about 2 wpi (Fig. 5a). Newly developed upper trifoliate leaves show a highly uniform chlorosis (Fig. 5b). Non-infected or wild-type ALSV (wtALSV) infected seedlings do not show photo-bleaching (Fig. 5a and b). SoyPDS-ALSV is detected from the photo-bleached area in infected leaves by tissue-blot hybridization analysis (Fig. 5c).

Fig. 5. (**a**) *Left* is a non-infected soybean, and *right* is a soyPDS-ALSV infected soybean, both at 14 days post inoculation. (**b**) The photo-bleached phenotype of a soyPDS-ALSV infected soybeans is observed on a new developed leaves at later growth stages. (**c**) Tissue-blot hybridization analysis of soyPDS-ALSV infected soybean leaves. Upper panel shows each leaf of a soyPDS-ALSV infected soybean. Lower panel shows ALSV signal (*black region*) by tissue blot hybridization. soyPDS-ALSV is infected systemically and ALSV signal is strongly detected from the region showing photo-bleaching.

4. Notes

1. General molecular techniques according to the standard protocol (17) were performed, for example, construction of ALSV vectors, *E. coli* transformation and agarose gel electrophoresis.
2. If more ALSV vector infected *C. quinoa* is needed, leaf tissues infected with ALSV vector (0.1 g) are homogenized in 0.3 mL of ALSV inoculation buffer (0.1 M Tris–HCl (pH 7.5), 0.1 M NaCl, 0.05 M $MgCl_2 \cdot 6H_2O$. Sterilize by autoclave. Store at 4°C.). A cotton swab is soaked with leaf extracts and rubbed on the surface of *C. quinoa* leaves (eight leaf stage) (fifth to eights true leaves) which are dusted with 600-mesh carborundum. A surface of inoculated leaves is flushed with a sufficient amount of water to remove carborundum. The symptoms consisting of chlorotic spots and mosaic appeared in the upper uninoculated leaves at about 7–10 days post inoculation (dpi). *C. quinoa* leaves with symptoms are collected and stored at −80°C.
3. RNA-coated gold particles prepared by this method can be used for inoculation to cotyledons of soybean using the PDS-1000/He Particle Delivery System which is another biolistic inoculation system (unpublished).
4. At this point, please confirm that there are no clumps of gold particles in the suspension.

Acknowledgements

We thank S. Sasaki for his technical support. This work was supported in part by Grant-in-Aids for Research and Development Projects for Application in Promoting New Policy of Agriculture, Forestry and Fisheries from the Ministry of Agriculture, Forestry and Fisheries, KAKENHI (no. 20380025) from the Ministry of Education, Culture, Sports, Science and Technology of Japan, and Programme for Promotion of Basic and Applied Researches for Innovations in Bio-oriented Industry.

References

1. Lu R et al (2003) Virus-induced gene silencing in plants. Methods 30:296–303
2. Waterhouse PM, Helliwell CA (2002) Exploring plant genomes by RNA-induced gene silencing. Nat Rev Genet 4:29–38
3. Burch-Smith TM et al (2004) Applications and advantages of virus-induced gene silencing for gene function studies in plants. Plant J 39:734–746
4. Bruun-Rasmussen M et al (2007) Stability of Barley stripe mosaic virus-induced gene silencing

in barley. Mol Plant Microbe Interact 20: 1323–1331

5. Dong Y et al (2007) A ligation-independent cloning tobacco rattle virus vector for high-throughput virus-induced gene silencing identifies roles for NbMADS4-1 and -2 in floral development. Plant Physiol 145:1161–1170
6. Gossele V et al (2002) SVISS—a novel transient gene silencing system for gene function discovery and validation in tobacco plants. Plant J 32:859–866
7. Purkayastha A, Dasgupta I (2009) Virus-induced gene silencing: a versatile tool for discovery of gene functions in plants. Plant Physiol Biochem 47:967–976
8. Li C et al (2000) Nucleotide sequence and genome organization of Apple latent spherical virus: a new virus classified into the family Comoviridae. J Gen Virol 81:541–547
9. Li C et al (2004) Stable expression of foreign proteins in herbaceous and apple plants using Apple latent spherical virus RNA2 vectors. Arch Virol 149:1541–1558
10. Takahashi T, Yoshikawa N (2008) Analysis of cell-to-cell and long-distance movement of Apple latent spherical virus in infected plants using green, cyan, and yellow fluorescent proteins. In: Forester G, Johansen E, Hong Y, Nagy P (eds) Plant virology protocols (series methods in molecular biology, vol. 451). Humana Press, USA, pp 545–554
11. Yaegashi H et al (2007) Characterization of virus-induced gene silencing in tobacco plants infected with Apple latent spherical virus. Arch Virol 152:1839–1849
12. Igarashi A et al (2009) Apple latent spherical virus vectors for reliable and effective virus-induced gene silencing among a broad range of plants infecting tobacco, tomato, *Arabidopsis thaliana*, cucurbits, and legumes. Virology 386:407–416
13. Yamagishi N, Yoshikawa N (2009) Virus-induced gene silencing in soybean seeds and the emergence of soybean plants with Apple latent spherical virus vectors. Plant Mol Biol 71:15–24
14. Yamagishi N et al (2011) Promotion of flowering and reduction of a generation time in apple seedlings by ectopical expression of the *Arabidopsis thaliana* FT gene using the Apple latent spherical virus vector. Plant Mol Biol 75:193–204
15. Sasaki S, Yamagishi N, Yoshikawa N (2011) Efficient virus-induced gene silencing in apple, pear and Japanese pear using Apple latent spherical virus vectors. Plant Methods 7:15. doi:10.1186/1746-4811-7-15
16. Yamagishi N, Sasaki S, Yoshikawa N (2010) Highly efficient inoculation method of apple viruses to apple seedlings. Julius-Kuhn-Archives 427:226–229. http://pub.jki.bund.de/index.php/JKA/article/view/443/1496
17. Sambrook J, Russell DW (2001) Molecular cloning: a laboratory manual, 3rd edn. Cols Spring Harbor Laboratory Press, Cold Spring Harbor, New York, NY

Chapter 14

VIGS: A Tool to Study Fruit Development in *Solanum lycopersicum*

Josefina-Patricia Fernandez-Moreno, Diego Orzaez, and Antonio Granell

Abstract

A visually traceable system for fast analysis of gene functions based on Fruit-VIGS methodology is described. In our system, the anthocyanin accumulation from purple transgenic tomato lines provides the appropriate background for fruit-specific gene silencing. The tomato Del/Ros1 background ectopically express *Delila (Del)* and *Rosea1 (Ros1)* transgenes under the control of fruit ripening E8 promoter, activating specifically anthocyanin biosynthesis during tomato fruit ripening. The Virus-Induced Gene Silencing (VIGS) of *Delila* and *Rosea1* produces a color change in the silenced area easily identifiable. *Del/Ros1* VIGS is achieved by agroinjection of an infective clone of *Tobacco Rattle Virus* (pTRV1 and pTRV2 binary plasmids) directly into the tomato fruit. The infective clone contains a small fragment of *Del* and *Ros1* coding regions (named DR module). The co-silencing of reporter *Del/Ros1* genes and a gene of interest (GOI) in the same region enables us to identify the precise region where silencing is occurring. The function of the GOI is established by comparing silenced sectors of fruits where both GOI and reporter DR genes have been silenced with fruits in which only the reporter DR genes have been silenced. The Gateway vector pTRV2_DR_GW was developed to facilitate the cloning of different GOIs together with DR genes. Our tool is particularly useful to study genes involved in metabolic processes during fruit ripening, which by themselves would not produce a visual phenotype.

Key words: Virus-Induced Gene Silencing, Tomato fruit, Agroinjection, Tobacco Rattle Virus, Gateway, pTRV2_Del/Ros1_GW, Co-silencing, Anthocyanin, Gene function.

1. Introduction

Virus-Induced Gene Silencing (VIGS) is a technique based on RNA-mediated antiviral plant defense that has been used to analyze gene function in plants (1–4). A fragment of a plant gene of interest (GOI) is inserted into the recombinant viral genome used for infection. The specific degradation of endogenous GOI's mRNA is the result of plant antiviral defense and produces the silencing of the endogenous GOI (5). VIGS presents multiple advantages when compared to other loss-of-function techniques

Annette Becker (ed.), *Virus-Induced Gene Silencing: Methods and Protocols*, Methods in Molecular Biology, vol. 975, DOI 10.1007/978-1-62703-278-0_14,

(3, 5–7) and therefore qualifies as an advantageous technique for reverse genetic studies.

Genomics project are generating an overwhelming amount of information, thanks to a more powerful RNA sequencing and array technologies. This is also happening with tomato, which is a particularly important crop, not only because its fruit contributes importantly to the human diet, but also because it is becoming a model crop species. To improve both the nutritional value and organoleptic features of this crop, first it is necessary to understand the genetic basis of the metabolic pathways that operate during fruit ripening processes (6). The easy way in which we can obtain genomic information contrasts with our current lack of understanding about the function of many genes in the genome. In many fruit crops, one way to investigate gene function is altering its expression by stable transformation. Different techniques can be used for that objective, but are often cumbersome and lengthy taking several months or years. A rapid and high-throughput method is required, which allows both to analyze the enormous amount of data from genomic projects and to link gene functions to phenotypes (6). VIGS technique can be successfully applied in tomato fruit for that purpose (3, 6, 8–10). The use of *Agrobacterium tumefaciens* as a vehicle for transfection is the common way to introduce effectively viral-modified vectors for VIGS approaches. We developed a new VIGS methodology in fruits named "Agroinjection" (Fig. 2) which introduces *Agrobacterium* suspension into tomato fruit tissues by stylar apex (8). This method speeds up the experimental procedures and confines the VIGS signal into the fruit, allowing to increase the throughput of VIGS by "one organ-one biological replicate" approaches (8).

Different viruses have been used as suitable VIGS vectors. *Tobacco Rattle Virus (*TRV) was described as a VIGS vector one decade ago (1), and since then it has been one of the most widely used (4). TRV-based vectors for VIGS approaches consist in pTRV1 and pTRV2 binary plasmids. GOIs are cloned into pTRV2 plasmid by digestion/ligation cloning or into pTRV2_GW vector by Gateway recombination (11). In tomato fruits, TRV-based vectors normally produce partial VIGS penetration and patchy tissue distribution as a result from partial and highly variable silencing from fruit to fruit (6). This causes serious limitation for its use in the investigation of gene loss-of-function that yields nonvisual phenotypes (6). An internal reference that monitors the levels of silencing was developed to overcome these limitations and increase the sensitivity of downstream analysis, allowing the dissection of silenced from non-silenced tissues (6). In our system, the anthocyanin accumulation in purple transgenic tomato lines provides the appropriate background for fruit-specific gene silencing. These lines were obtained in Dr. C. Martin's group (12) by ectopically expression of Delila (Del) and Rosea1 (Ros1) genes (two transcription factors that activate the anthocyanin branch of flavonoid biosynthesis

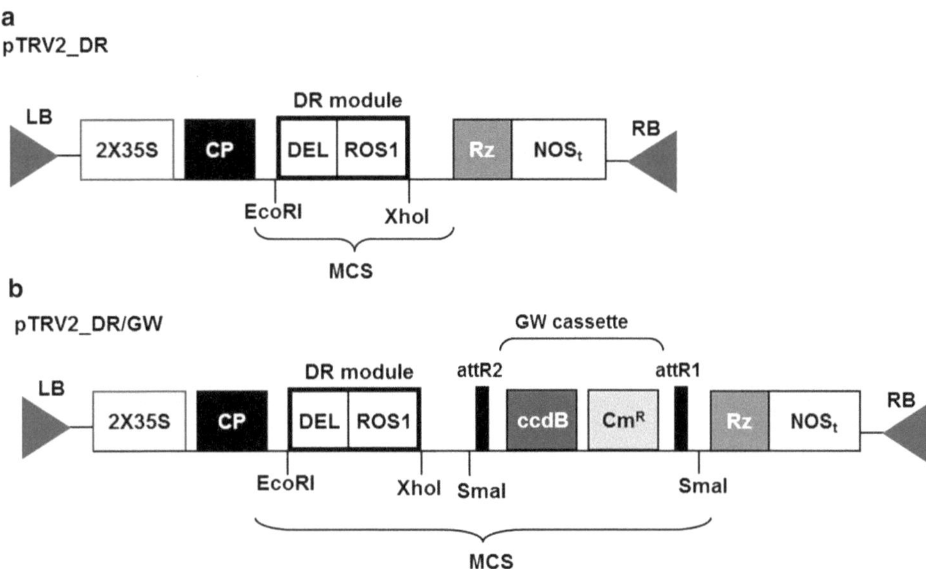

Fig. 1. TRV-based vectors for VIGS that could be used with this protocol: (**a**) Standar pTRV2 for traditional cloning in the MCS and containing the DEL ROS1 module for monitoring the regions of silencing; (**b**) Gateway adapted pTRV2_DR/GW vector with the DEL ROS1 monitoring module and the GW casette for cloning the gene of interest.

pathway in *Antirrhinum majus* flowers (13)) under the control of tomato E8 promoter. This resulted in the activation of anthocyanin biosynthesis specifically during tomato fruit ripening (6, 12). The silencing of both *Del* and *Ros1* genes, using small fragments of their coding regions (named reporter DR genes) by pTRV2_DR expression vector (developed in our laboratory, Fig.1a; (6, 9)), results in the lack of anthocyanin production. As reporter DR genes silencing involves the blockage of a pathway not normally active in tomato fruit, the lack of anthocyanin accumulation produces red silenced sectors that present similar characteristics in metabolism and development as "wild type" tomato fruit (6). To facilitate the dissection of silenced tissues and to increase the yield of silenced areas for downstream analysis, we transferred the *Del* and *Ros1* transgenes from cherry-type MicroTom (12) to a large globe-type MoneyMaker tomato background (Fig. 3a) by standard crossing and selection (see Subheading 2.1, item 1; (6)). The integration of DR-reporter module and GOI in the same viral genome (pTRV2_DR_GOI VIGS vector) is required for an efficient co-silencing of both the reporter module and target gene in the same tissue area (co-silencing in *tandem*; example in Fig. 3b, c) (6). To facilitate high-throughput tandem cloning of subsequent GOIs, we modified pTRV2_DR vector into pTRV2_DR_GW vector (Fig. 1b) by the introduction of a Gateway recombination cassette (6). This system has proved to be particularly useful for the analysis of genes of unknown function involved in different stages of fruit ripening, especially of genes associated with different branches of metabolism in fruit (6, 10).

In the present chapter we describe the methodology of Fruit-VIGS based on anthocyanin accumulation in tomato fruit, as a tool to study the gene function during ripening stages related with quantitative characters, as secondary metabolites in tomato fruit. Details on plant cultivation and maintenance are explained. Some recommendations on silenced areas harvesting and their analysis are provided, too. At the end of the chapter, we present some future perspectives to use this tool with different reporter genes and different promoters aimed to study genes involved in different developmental stages of tomato fruit.

2. Materials

2.1. Plant Material

1. Globe-type purple tomatoes (Solanum lycopersicum) were obtained by crossing *Del/Ros1* MicroTom N line (T2 homozygous generation from Micro Tom plants transformed with *Delila* and *Rosea1* cDNAs under the control of the E8 ripening-specific promoter; (12)) with wild-type MoneyMaker plants (6). Segregating sibling lines were selfed and selected through to the F7 generation. Selection was based on globe-type fruit, smooth leaves, indeterminate growth, and best fruit VIGS response (6).
2. Plants were grown in a greenhouse supplemented with artificial light from mercury vapor lamps (OSRAM) of 400ω (PHILILPS HDK) 400HPI®N (96 μmol m^{-2} s^{-2}(14)).
3. Plants were irrigated four times per day with a HOAGLAND N°1 nutritive solution supplemented with oligo elements by automatic dripping irrigation system (14).

2.2. Cloning Procedures and Vectors Construction

Gateway technology (www.invitrogen.com) has been used to generate the different VIGS vectors following the manufacturer's instructions.

1. For amplification: Advantage® 2 DNA Polymerase Mix (Clontech, Mountain View, CA, USA; see Note 1), specific forward and reverse primers (10 mM each one), dNTPs mixture (10 mM each dNTP), and sterilized water.
2. For DNA cleanup: QUIAquick® PCR Purification kit (Qiagen, Valencia, CA, USA) and sterilized water; or appropriate percentage of agarose in TAE 1× (40 mM Tris–Acetate and 1 mM Na_2EDTA) and QUIAEXII® Gel Extraction kit (Qiagen, Valencia, CA, USA).
3. NanoDrop spectrophotometer (NanoDrop ND-100 Spectrophotometer; Thermo Fisher Scientific Inc., USA). Follow the manufacturer's instructions in each case.

4. For cloning in pDONOR to generate a pENTR clone: pCR®8/GW/TOPO® TA Cloning® Kit (Invitrogen, Carlsbad, CA, USA) and freshly purified PCR product. Follow the manufacturer's instructions.
5. For final generation of a VIGS Vector (Expression clones): TRV-based silencing vectors pTRV1 and pTRV2 were provided by Prof. Dinesh Kumar (11); pTRV2_DR and pTRV2_DR_GW VIGS vectors were generated in our group (Fig. 1, (6)).
6. Gateway® LR Clonase™ II Enzyme Mix (Invitrogen, Carlsbad, CA, USA).
7. For *E. coli* transformation, use One Shot®TOP10 or One Shot®Mach1™T1R chemically competent *E. coli* kit (Invitrogen, Carlsbad, CA, USA).
8. Sterile LB liquid medium and solid LB agar plates containing 50 μg/mL Spectinomycin in case of Entry clones (TA-Cloning) or 50 μg/mL Kanamycin for Expression clones (LR reaction). 37°C growing chamber.
9. For *E. coli* DNA plasmid extraction use Plasmid Mini Kit I E.Z.N.A. (Omega Biotek, Doraville, GA, USA).
10. For *E. coli* colony glycerol stocks: in a sterile Eppendorf tube mix 700 μL of fresh liquid *E. coli* culture and 300 μL of 50% sterile glycerol. Freeze it quickly in liquid N2 and store it at −80°C.

2.3. Agrobacterium Transformation and Agroinjection Suspension Preparation

1. For *Agrobacterium* transformation: Eppendorf tubes containing 40 μL of electrocompetent *Agrobacterium* C58 cells stored at −80°C (see Note 2).
2. Electroporator (Bio-Rad, gene-pulser 165-2077) + 1 mm electroporation cuvettes (Bio-Rad Laboratories, CA, USA).
3. 15 mL plastic tubes containing 250 μL of S.O.C. medium (2% tryptone, 0.5% yeast extract, 10 mM NaCl, 2.5 mM KCl, 10 mM $MgCl_2$, 10 mM $MgSO_4$, and 20 mM glucose).
4. Sterile LB liquid medium and solid LB agar plates containing 50 μg/mL kanamycin and 50 μg/mL rifampicin antibiotics
5. Growing chamber set at 28°C.
6. For *Agrobacterium* DNA plasmid extraction use QUIAprep® Miniprep Kit (Qiagen, Valencia, CA, USA).
7. For *Agrobacterium* glycerol stocks: in a sterile Eppendorf tube mix 700 μL of fresh liquid *Agrobacterium* culture and 300 μL of 50% sterile glycerol. Freeze it quickly in liquid N2 and store it at −80°C.
8. For *Agrobacterium* C58 cultures and subcultures for agroinjection: 15 mL plastic tubes.
9. MES infiltration buffer: 10 mM MES (Sigma-Aldrich, MO, USA; see Note 3), 10 mM MgCl (see Note 3), 200 μM

acetosyringone (Sigma-Aldrich, MO, USA; see Notes 3 and 4). Rotating and swaying mixer (CAT RM-5).

10. Spectrophotometer (UV/VIS Spectrophotometer SP8001, DINKO) set at a wavelength of 600 nm and transparent plastic cuvettes.

2.4. Fruit Agroinjection

1. Tomato *Del/Ros1* fruits at Mature Green (MG) stage, 30–35 days post-anthesis (dpa).
2. Sterile 1 mL Plastipak needle syringes (25 GA 5/8 IN, needle: 0.5 × 16 mm, BD Plastipak™).

2.5. Dissection and Collection of Silenced Sectors

1. For silenced sectors dissection: Glass board to dissect the fruit with sharp knife (see Note 5).
2. For silenced sectors collection: Plastic screwed cap tubes (25 mL) to store the samples at −80°C. Liquid Nitrogen (N2) in a suitable container.
3. For silenced sectors crushing: Thermal-cover mortar, metallic little spoon, thermal gloves, and protective glasses.

2.6. Evaluation of GOI Silencing

1. A suitable RNA extraction method for tomato fruits (15).
2. SuperScript TMFirst-Strand Synthesis System for RT-PCR (Invitrogen, Carlsbad, CA, USA). Follow the manufacturer's instructions.
3. Power SYBR® Green PCR Master Mix and RT-PCR (Applied Biosystems, Madrid, CA, USA) and 7500 Fast Real-Time PCR system (Applied Biosystems, Madrid, CA, USA).

3. Methods

3.1. Plant Cultivation and Maintenance

1. Estimation of the plant numbers required for the experiment. For that, determine the following: (a) how many genes will be silenced; (b) how many individual silencing constructions will be generated; and (c) how much material will be necessary for downstream analysis (see Note 6). For each construction we use three plants with 20 fruits each, ten of them for control DR-silencing and the other ten for GOI-DR silencing (see Note 7).
2. Sowing and seedlings. Sow more seeds than plants will be required and keep some extra seedlings to allow selection of best-performing plants (see Note 8).
3. Pruning and labeling. Remove secondary buds to give structure to the plant (a main axis with first lateral branches only) and to increase reproductive vigor (see Note 9). During flowering

period, flower unloading and labeling are required. Keep six flowers per truss and label them with the anthesis date. When small fruits appear, keep six or seven floral trusses and remove the rest. Later, the excess of fruit per truss should be removed. But before that, make sure that at least four of them have reached Inmature Green (IMG) stage (see Note 10). When the fruits get to Mature Green (MG) stage, select which ones will be used for DR control silencing and which others will be used for GOI-DR co-silencing. Then label them appropriately (see Note 11) before agroinjection.

3.2. Cloning Procedures and Vector Construction

1. Design the primers for a specific GOI region. Optimal GOI regions are between 100 and 500 bp in length (see Note 12).
2. GOI sequence amplification by PCR reaction using Advantage® 2 DNA Polymerase Mix. PCR conditions will depend on fragment length and primers Tm.
3. Amplified DNA fragment purification. If the PCR yields a single product, purify the PCR reaction using QUIAquick®PCR Purification kit, and elute it in 50 μL of volume. If the PCR yields several products, put the complete PCR reaction on an electrophoresis gel with appropriate percentage of agarose. Separate the product of interest and excise it from the gel. Purify it using QUIAEXII® Gel Extraction Kit, and elute in 30 μL of volume. In both cases, quantify the DNA by NanoDrop spectrophotometry.
4. Ligation to obtain the pCR8_GOI entry clone. Prepare as many 1.5 mL Eppendorf tubes as amplified GOIs. Prepare the TA-Cloning reaction following the pCR®8/GW/TOPO® TA Cloning® Kit manufacturer's instructions. Incubate the ligation 1 h at room temperature.
5. Transformation of entry clone in *E. coli*. Take from –80°C a tube of chemically competent *E. coli* cells per each Entry clone to be generated (One Shot®TOP10 or One Shot®Mach1TMT1R (see Subheading 2.2, item 5)). Thaw competent cells on ice. Add 2 μL of pCR8-GOI ligation reaction to each tube, incubate without shaking and perform transformation following manufacturer's instruction (see Note 13). Collect 50 μL from the bacterial culture and spread in a solid LB plate containing 50 μg/mL spectinomycin (see Note 14). Incubate the plates overnight at 37°C.
6. Validation of entry clone. Pick 4–6 colonies into 3 mL of liquid LB media with 50 μL/mL spectinomycin using toothpicks in sterile conditions. Allow them to grow overnight at 37°C. Isolate the plasmid DNA using the E.Z.N.A.® Plasmid Mini Kit I, and elute in 50 μL of volume. Validate the pCR8_GOI entry clone by restriction analysis (see Note 15) and by sequencing using M13 forward and reverse primers. Generate

a glycerol stock (see Subheading 2.2, item 7) with a positive colony previously validated.

7. Generation of final expression vector. For each LR reaction, mix in a 1.5 mL sterile Eppendorf: 50–150 ng Entry clone (pCR8-GOI) (1–7 μL), 150 ng/μL (1 μL) of pTRV2_DR_GW destination vector, and TE buffer pH 8.0 up to 8 μL. Follow the instructions from Gateway® LR ClonaseTM II Enzyme Mix manufacturer's instructions.
8. Transformation of final expression vector. Proceed as Subheading 3.2, step 5 but using 1 μL of LR reaction and 50 μg/mL kanamycin in selective LB plates.
9. Validation of final expression vector. After overnight incubation, pick 4–6 colonies and proceed as Subheading 3.2, step 6. Validate the expression vector by restriction analysis and sequencing (see Note 16). Finally, generate a glycerol stock (see Subheading 3.2, step 6).

3.3. Agrobacterium Transformation and Agroinjection Suspension Preparation

1. *Agrobacterium* transformation. Take a sterile Eppendorf containing 40 μL of *Agrobacterium* C58 electrocompetent cells per GOI and thaw them on ice. Add to each tube 1 μL of a positive *E. coli* plasmid miniprep from Subheading 3.2, step 9. Electroporate the samples at 1.5 V. Add 250 mL of SOC medium and incubate them in 15 mL plastic cap tube for 2 h shaking (150–200 rpm) at 28°C. Then, collect them by spin at 12,000 × *g* in a micro-centrifuge. Remove the supernatant, leaving approximately 100 μL. Resuspend the cells and spread them on selective LB plates containing 50 μg/mL of both kanamycin and rifampicin. Incubate for at least 48 h at 28°C.
2. Validation of *Agrobacterium* clones. Pick 4 colonies from selective plates in sterile conditions (as Subheading 3.2, step 6) and allow them grow in 5 mL of liquid LB media containing kanamycin and rifampicin (50 μg/mL) at 28°C for 48 h. Isolate the plasmid DNA by miniprep and validate by digestion with the suitable restriction enzyme (see Note 17). Generate a glycerol stock with a positive colony previously validated.
3. Culturing and subculturing of *Agrobacterium* clones. Grow pTRV1, pTRV2_DR, and each pTRV2_DR_GOI construct from frozen stocks individually in kanamycin and rifampicin (50 μg/mL) selective LB plates. Pick a colony of each LB plate and put them into a 50 mL plastic tube containing 5 mL of LB medium with kanamycin and rifampicin (50 μg/mL). Grow them shaking 48 h at 28°C. Based on the number of labeled MG fruit (Subheading 3.1, step 3), make an estimation of *Agrobacterium* suspension volume required for each final expression vector. The volume of agroinjection suspension mix (pTRV1:each pTRV2) varies depending on the fruit size, but

1 mL should be enough to infiltrate one MG Del/Ros1 MM fruit. For small fruits, use 0.5 mL of agroinjection suspension. Prepare a fresh pre-culture of *Agrobacterium*. Take 100 μL of each pre-cultures and inoculate a 50 mL plastic tube containing 5 mL LB medium with kanamycin and rifampicin (50 μg/mL) (Fig. 2a). Grow by shaking overnight at 28°C.

4. Agroinjection suspension preparation. Collect the *Agrobacterium* cells by centrifugation at 3,000 × *g* for 15 min. Discard the supernatant by inversion. Prepare acetosyringone solution and add it to MES infiltration buffer. Protect it from light wrapping the bottle with aluminum foil. Resuspend the pellet with the cells in 15 mL MES infiltration buffer and vortex it (see Note 18) to produce the agroinjection suspension. Wrap the plastic tubes in aluminum foil, too. Incubate them at room temperature with gently agitation (20 rpm) in a rotating and swaying mixer for at least 2 h. Check the optical density (OD) at 600 nm wavelength of each suspension and dilute them adding more MES infiltration buffer to reach 0.05 OD (see Note 19). Prepare the agroinjection suspension by mixing 1:1 (volume–volume) the pTRV1 suspension with each pTRV2 suspensions, including pTRV2_DR control (Fig. 2a, b, see Note 20).
5. Fruit agroinjection. Use different sterile 1mL needled syringes for each agroinjection suspensions (containing pTRV1 and pTRV2_DR or PTRV1 and each pTRV2_DR/GOI vectors) and agroinject them into MG fruits (30–35 dpa) (2c, e). Agroinjection proceeds by inserting the needle about 3–4 mm into the fruit through the calyx region, between sepals and peduncle junction (carpopodium). This was found to be more efficient than injection through the stylar end as initially described (9). Inject the suspension carefully (see Note 21). The successful spread of agroinjection suspension into the fruit can be monitored by the color change observed in the fruit tissues from light to dark green. Agroinjection is finished when the fruit is fully infiltrated, and a few drops appear in the sepal hydatodes (Fig. 2f). Dry the drops on the fruit and keep the fruit surface clean. DR and GOI gene silencing can be observed 10 days after agroinjection, when the fruit reaches the breaker stage.

3.4. Dissection and Collection of Silenced Sectors

1. Harvesting. Fruits are harvested at different ripening stages depending on the particular interest of each study. Harvest DR- and DR/GOI-silenced fruits separately and keep their labels.
2. Dissection and collection. Rinse fruits with tap water and dry them. Separate silenced from non-silenced areas by cutting them with a sharp knife (Fig. 2d). For pericarp tissue studies, slice the fruit and discard seeds and gel. Sort out silenced from non-silenced areas (see Note 21), transfer them quickly to conveniently labeled screw cap tubes (25 mL) and hold in liquid N_2.

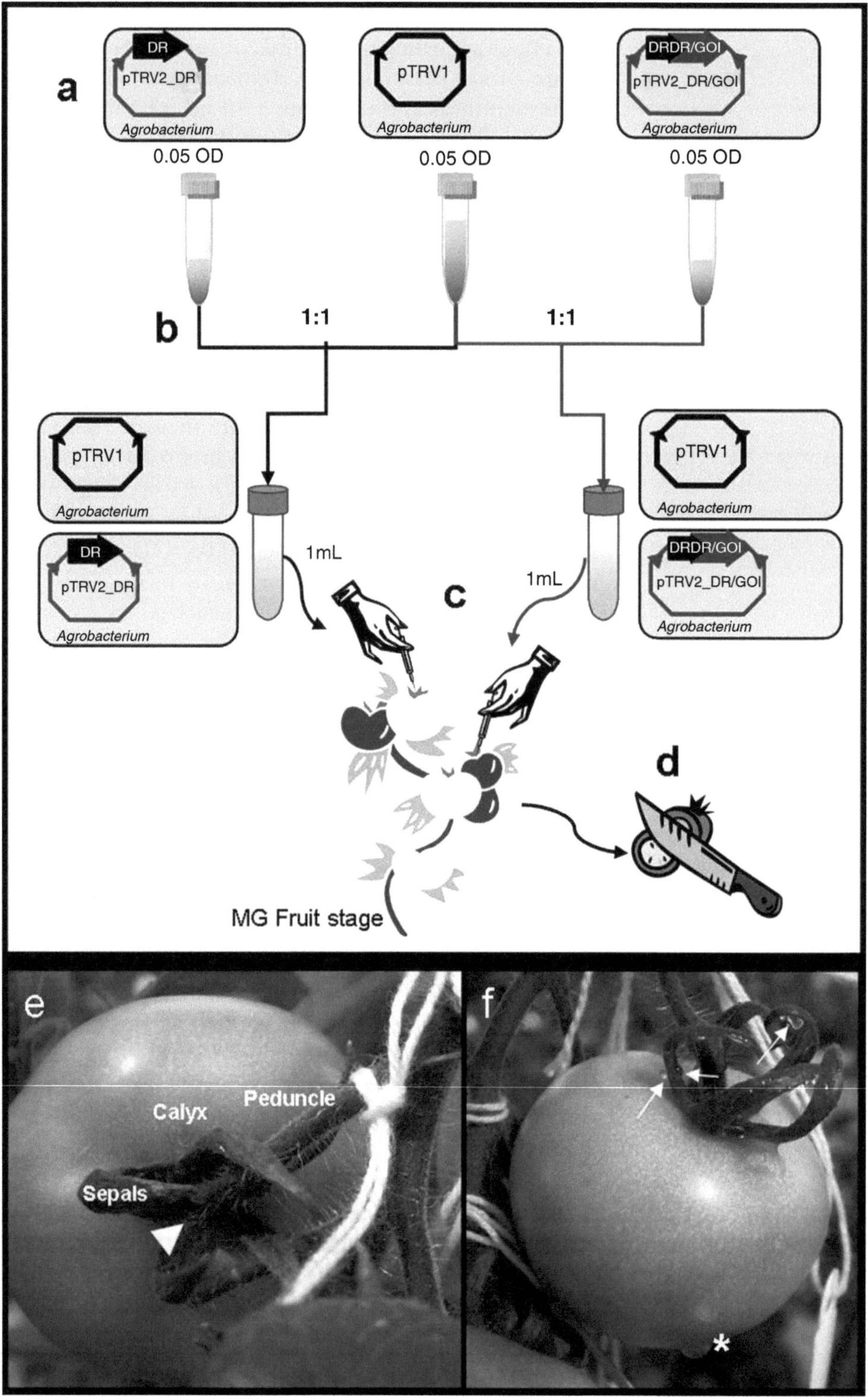

Fig. 2. Schematic of the Agrobacterium inoculum preparation for fruit agroinfiltration (**a-d**), and detail of the injection site in the carpopodium region (**e**) and gutattion emerging from the sepals and fruit after efficient injection.

Fig. 3. Some phenotypes obtained by fruit VIGS using the DelRos background: (**a**) non silenced DR fruit; (**b**) red sectors with DR silenced (DR_S) showing the wild type red fruit pheotype and non silenced sectors (DR_NS); (**c**) cosilencing of DR and PDS producing regions with no accumulation of both carotenoids and anthocyanins (DR/PDS_S) next to regions where no silencing is occurring DR/PDS_NS.

Store at −80°C. Grind the samples up in a mortar with liquid N_2 to obtain a fine frozen powder. In Fig. 3 we show an example of DR- and DR/GOI-silenced fruits in comparison with DR-non-silenced fruit. We used the *Phytoene desaturase (PDS)* gene as GOI because it provides a visual phenotype by itself, which gives us the opportunity to evaluate the co-silencing of the DR module and the PDS gene (6).

3.5. Evaluation of GOI Silencing

1. Extract the RNA from the silenced region using a suitable RNA isolation protocol (15).
2. Synthesize cDNA from isolated RNA with SuperScriptTMFirst-Strand Synthesis System for RT-PCR following the manufacturer's instructions.
3. Perform a relative quantification of transcript abundance e.g., in 7,500 Fast Real-Time PCR system (Applied Biosystems) using Power SYBR® Green PCR Master Mix and an established RT-PCR protocol.

3.6. Future Perspectives

The strategy used in purple *Del/Ros1* transgenic plants as background that provides a monitoring system for VIGS can be adapted to other developmental stages of tomato fruit by engineering. Different approaches are possible to achieve this objective: (a) to change the stage-specific promoter. The pENFRUIT collection developed in our laboratory can be used as promoter source for specific developmental stages (16); (b) to change the reporter gene. Other visual reporters easily traceable can be used, e.g., fluorescent DsRed protein. The *35S-DsRed* MM tomato transgenic line developed in our laboratory, works as a new and attractive monitoring VIGS strategy for early stage of fruit development (9); and (c) to combine both different specific-stage promoters with different reporter genes. In this way, complete developmental and ripening processes could be studied in tomato fruit.

4. Notes

1. We have used Advantage™ as a high fidelity polymerase which adds a single adenine overhang at 3 ends of each amplicon. It is necessary for TA cloning reaction.
2. In tomato fruit (*Solanum lycopersicum*) we observed that *Agrobacterium* C58 strain is more infective and efficient in transient silencing assays than LBA4404.
3. Stocks for MES infiltration buffer: (a) MES 100 mM (10×) pH 5,6. Dissolve in sterile water and adjust the pH with KOH 1M. Require sterilization by autoclaving or filtration; (b) Magnesium Chloride (MgCl2) 1M (100×). Dissolve in sterile water. Require sterilization by filtration; and (c) 200 mM acetosyringone solution. Dissolve 78.48 mg in 2 mL of dimethyl sulfoxide and filter-sterilize it. Split in 200 μL aliquots and store at -20°C. IMPORTANT: It is better to prepare the 200 mM stock acetosyringone on the same day you plan to use it.
4. Acetosyringone is photosensitive and it needs to be under darkness conditions. Tubes are wrapped with aluminum foil.
5. A sharp knife is important to avoid crushing the tissue.
6. Depending on many variables, such as the final expression vector size or environmental conditions, the yield of silenced sectors per fruit will be different. For example, in our experience we observed that expression vectors with larger sizes produce less silenced areas per fruit than vectors with shorter sizes. In these cases, we advise to use more plants per experiment in order to increase the amount of silenced samples.
7. Based on our experience, good results were obtained with 20 fruits per plant. We try to keep five trusses per plant, containing four fruits each. In each truss, two fruits are used to control silencing and other two fruits are used for DR/GOI co-silencing. With three plants per construction, enough material is obtained for most downstream analysis (e.g., metabolomics).
8. Even though our *Del/Ros1* MM background is a F7 generation, sometimes one or two plants with MT trait appear. They should be removed.

9. If pruning can affect the expression of your trait, do not prune the plants. If this is not the case, prune them every week until the last fruit is collected.
10. Put special attention at plant pruning after first fruit reach Mature Green (MG) stage: Remove every floral truss and secondary bud.
11. Label the MG fruits indicating date, silencing vectors agroinjected and write a code that contains fruit, truss, and plant number. IMPORTANT: keep this code for downstream analysis because it represents a unique biological replicate.
12. Selection of a specific GOI region is particularly important when working with gene families. Overlapping primer designs can be used to clone several gene fragments in *tandem* inside the same vector.
13. We routinely use *E. coli* (One Shot®TOP10 or One Shot®Mach1 TMT1R chemically competent *E. coli* kit) following the transformation conditions from pCR®8/GW manual.
14. When the transformation efficiency has not been optimal, we spread the rest of culture (150 μL) in a different plate for recover some colonies.
15. We use Vector NTI 10.3.0 (Jul 31, 2006© Invitrogen, Carlsbad, CA, USA) program to choose the suitable restriction enzymes and predict the expected sizes after plasmid DNA digestion. Proper enzymatic reaction conditions can be found at the enzyme manufacturer website. The pCR®8/GW/TOPO® TA Cloning® Kit, recommend *EcoRI* restriction enzyme because it releases the cloned GOI from the plasmid. Be careful with additional *EcoRI* digestion products which can be obtained as a result of internal *EcoRI* sites in your fragment.
16. We use Vector NTI 10.3.0 program, too, to select the best restriction enzyme. We commonly use *EcoRV* (Takara Bio Europe, France; Takara Bio Inc., Shiga, Japan).
17. To increase the yield of *Agrobacterium* plasmid DNA minipreps, collect the cells from 5 mL liquid LB medium and follow QUIAprep® Miniprep Kit procedures. Elute with 20 μL to obtain a more concentrated plasmid DNA preparation.
18. It is IMPORTANT to resuspend completely the cells in the MES infiltration buffer. Vortex them for around 1–2 min.
19. Usually, we make a 0.5 OD intermediate dilution for each agroinjection suspension in order to equalize them before reaching the final 0.05 OD.
20. In some fruits you may find a higher initial resistance to agroinjection than normal. In those cases try to find a more suitable position by changing the depth of needle insertion in the fruit.

21. Sometimes you may find it difficult to distinguish between silenced and non-silenced areas. Small patchy silencing or a gradual silencing aggravates the dissection of the sectors. If there are enough samples, discard those fruits. If not, try to select areas in patchy silenced fruit with high similarity to well-silenced fruit.

Acknowledgments

We are grateful to Prof. Dinesh Kumar who kindly provided us with TRV-based silencing vectors pTRV1 and pTRV2. We appreciate Prof. Cathie Martin for providing *Del/Ros1* transgenic tomato lines. The work described here was supported by BIO2008-034034 grant from the Spanish Ministry of Science and Technology, FPU fellowship from Spanish MICINN and EUSOL project from EU.

References

1. Ratcliff F, Martin-Hernandez AM, Baulcombe DC (2001) Technical Advance. Tobacco rattle virus as a vector for analysis of gene function by silencing. Plant J 25:237–245
2. Lu R, Martin-Hernandez AM, Peart JR et al (2003) Virus-induced gene silencing in plants. Methods (San Diego, Calif) 30:296–303
3. Fu DQ, Zhu BZ, Zhu HL et al (2005) Virus-induced gene silencing in tomato fruit. Plant J 43:299–308
4. Shao Y, Zhu HL, Tian HQ, Wang XG et al (2008) Virus-Induce Gene Silencing in Plant Species. Rus J Plant Physiol 55:168–174
5. Burch-Smith TM, Anderson JC, Martin GB et al (2004) Applications and advantages of virus-induced gene silencing for gene function studies in plants. Plant J 39:734–746
6. Orzaez D, Medina A, Torre S et al (2009) A visual reporter system for virus-induced gene silencing in tomato fruit based on anthocyanin accumulation. Plant Physiol 150:1122–1134
7. Unver T, Budak H (2009) Virus-induced gene silencing, a post transcriptional gene silencing method. Int J Plant Genomic 2009:198–680
8. Orzaez D, Mirabel S, Wieland WH et al (2006) Agroinjection of tomato fruits. A tool for rapid functional analysis of transgenes directly in fruit. Plant physiol 140:3–11
9. Orzaez D, Granell A (2009) Reverse genetics and transient gene expression in fleshy fruits: overcoming plant stable transformation. Plant Signal Behav 4:864–867
10. Ballester AR, Molthoff J, de Vos R et al (2010) Biochemical and molecular analysis of pink tomatoes: deregulated expression of the gene encoding transcription factor SlMYB12 leads to pink tomato fruit color. Plant Physiol 152:71–84
11. Liu Y, Schiff M, Dinesh-Kumar SP (2002) Virus-induced gene silencing in tomato. Plant J 31:777–786
12. Butelli E, Titta L, Giorgio M et al (2008) Enrichment of tomato fruit with health-promoting anthocyanins by expression of select transcription factors. Nat Biotechnol 26:1301–1308
13. Davies KM, Marshall GB, Marie Bradley J, Schwinn KE et al (2006) Characterisation of aurone biosynthesis in Antirrhinum majus. Physiol Plant 128:593–603
14. Marti C, Orzaez D, Ellul P et al (2007) Silencing of DELLA induces facultative parthenocarpy in tomato fruits. Plant J 52:865–876
15. Bugos RC, Chiang VL, Zhang XH, Campbell ER et al (1995) RNA isolation from plant tissues recalcitrant to extraction in guanidine. Biotechniques 19:734–744
16. Estornell LH, Orzaez D, Lopez-Pena L et al (2009) A multisite gateway-based toolkit for targeted gene expression and hairpin RNA silencing in tomato fruits. Plant Biotechnol J 7:298–309

Chapter 15

A Protocol for VIGS in *Arabidopsis Thaliana* Using a *One-Step* TYMV-Derived Vector

Isabelle Jupin

Abstract

Virus-induced gene silencing (VIGS) is an important tool for the analysis of gene function in plants, which can be adapted for high-throughput functional genomics in model plant species such as *Arabidopsis thaliana*.

Here we describe the use of the *Turnip yellow mosaic virus* (TYMV)-derived vector pTY-S that has the ability to induce VIGS in *Arabidopsis thaliana*. This vector harbors a cDNA copy of the viral genome, in which a unique *Sna*BI restriction site has been engineered. This site allows the cloning of 80 bp synthetic oligonucleotides corresponding to inverted-repeat fragments of the target gene while retaining the ability of the virus to move systemically. Silencing requires plants to be simply inoculated by abrasion with a few micrograms of intact plasmid DNA, thus precluding the need for in vitro transcription, biolistic, or agroinoculation procedures. This *one-step* TYMV-based VIGS system is therefore simple to use, cost-effective, and highly consistent, which are important parameters to consider towards the development of high-throughput infection procedures. Another important characteristic of this viral vector is its capacity to infect and induce silencing in meristem tissues.

Key words: TYMV, VIGS vector, Gene silencing, *Arabidopsis thaliana*, Infectious plasmid, Rub-inoculation

1. Introduction

TYMV, the type member of the genus *Tymovirus*, is a monopartite positive-strand RNA virus that infects many Brassicaceae, including Arabidopsis (1, 2). Its genome (Fig. 1a) consists of one RNA molecule of 6318 nt that encodes three open reading frames. The 69 K protein serves as the viral movement protein and silencing suppressor (3, 4), and the 206 K protein is required for viral replication (5, 6). The structural 20-kDa viral coat protein (CP) is translated from a subgenomic RNA that is produced during viral infection (7).

Annette Becker (ed.), *Virus-Induced Gene Silencing: Methods and Protocols*, Methods in Molecular Biology, vol. 975, DOI 10.1007/978-1-62703-278-0_15,

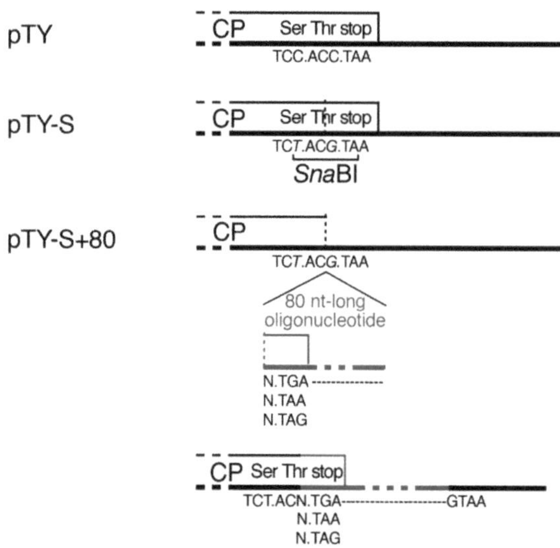

Fig. 1. Genomic organization of TYMV and derived vectors. (**a**) Genomic organization of TYMV RNA. The open reading frames are indicated by rectangles and correspond to the 69 K RNAi suppressor and movement protein, the 206 K replication polyprotein, and the coat protein (CP). The subgenomic promoter is indicated by an arrow. Details of the CP gene and protein sequences are shown below. (**b**) Details of the CP coding sequence in the TYMV-derived vectors, highlighting the requirement for an in-frame CP stop codon in the inserted 80 nt oligonucleotide.

Engineering TYMV into a VIGS vector first required to identify an insertion site in the viral genome where target gene sequences could be cloned. Because all three viral proteins are required for virus infection, we thus thought to introduce foreign sequences within the 3′ noncoding region of the viral genome. For that purpose, a unique *Sna*BI restriction site was engineered within the viral CP ORF, immediately upstream of the stop codon (Fig. 1b). Foreign sequences can be inserted in this site while maintaining virus infectivity, provided an in-frame CP stop codon is reconstituted (Fig. 1b) and that the inserted sequence does not exceed 80 nt in length (8). This small size therefore allows introduction of foreign sequences in the viral genome by direct cloning of double-stranded synthetic oligonucleotides. Interestingly, viruses with such an 80 nt insertion displayed reduced symptom severity (8).

Fig. 2. VIGS of the *PDS* gene using plasmid inoculation. Arabidopsis plants were inoculated with water (mock), the empty plasmid pTY-S or plasmid pTY-PDS-IR bearing an inverted-repeat sequence targeting the *phytoene desaturase* (*PDS*) gene. Photographs of cauline leaves and stems (**a–d**) and flowers and siliques (e-f) were taken 5–6 and 6–8 weeks post-inoculation, respectively. The *white* areas are caused by photobleaching due to silencing of the *PDS* gene (Reproduced from (8) with permission from John Wiley and Sons).

RNA silencing is ubiquitously triggered by double-stranded (ds) RNA molecules (9), and insertion of an inverted-repeat fragment of a target gene within the TYMV-derived VIGS vector successfully led to its gene silencing in *Arabidopsis* (8).

A further improvement of the technology was the development of a DNA-based version of this TYMV-derived VIGS vector. For that purpose, a cDNA copy of the modified viral genome was inserted into an expression cassette, positioned between the cauliflower mosaic virus 35S promoter and terminator. The resulting plasmid (pTY-S) was found to be highly infectious when miniprep-quality circular DNA was inoculated to plants simply by rub-inoculation in the presence of an abrasive, without the requirement of biolistic or agroinoculation procedures (8). Interestingly, this plasmid retained its ability to induce VIGS, and robust and reliable gene silencing in *Arabidopsis* was easily obtained by abrasion with the intact plasmid DNA (Fig. 2). This vector also had the ability to silence genes expressed in meristems, although not as efficiently as in vegetative tissues (8).

In this chapter, we describe the methods for conducting a VIGS experiment using the TYMV-derived vector pTY-S.

Obtaining a VIGS vector for a specific target gene first requires to design an 80 bp synthetic oligonucleotide with an inverted-repeat fragment of the gene of interest that reconstitutes an in-frame stop codon within the CP ORF. This oligonucleotide is then cloned within the *Sna*BI restriction site of pTY-S using standard molecular biology procedures. To initiate VIGS, the resulting plasmid DNA is then introduced into *Arabidopsis* leaves by rub-inoculation, using celite as an abrasive.

2. Materials

2.1. Plasmid Constructs

It is expected that standard molecular biology laboratory equipment are available. This includes the access to a heating block (or PCR machine), microbiology incubators at 37 and 30°C, microbiology shakers at 37 and 30°C, a low-speed centrifuge, a refrigerated centrifuge for Eppendorf tubes, a fume hood, an autoclave, –20 and –80°C freezers, a vortex, a water bath, a UV transilluminator, and a UV spectrophotometer (or NanoDrop, Peqlab) equipment.

More specific materials required consist in

1. Plasmid pTY-S.
2. Custom DNA oligonucleotide, 0.025 μmol synthesis scale, desalted.
3. Oligonucleotide hybridization buffer: 10 mM Tris–HCl (pH 7.5) and 50 mM NaCl.
4. Restriction enzymes *Sna*BI, *Eco*RI, and *Sma*I and their corresponding buffer (provided by their supplier).
5. T4 DNA ligase and its corresponding buffer (provided by its supplier).
6. Phenol/chloroform/isoamylic alcohol (25:24:1).
7. 96% ethanol.
8. 70% ethanol.
9. Purified and deionized (Milli-Q grade) sterile water.
10. Agarose (Molecular Biology grade).
11. Horizontal electrophoresis tank for agarose gels.
12. TBE buffer: 89 mM Tris base, 89 mM Boric acid, and EDTA 2 mM.
13. Ethidium bromide (0.625 mg/ml solution in water) or Sybr Green solution.
14. DNA molecular weight markers.
15. *Escherichia coli* (DH5α strain) competent cells (10^6 cfu/μg).
16. LB broth: 10 g Bacto-tryptone, 5 g yeast extract, and 10 g NaCl, adjust volume to 1 L with water. Autoclave and allow to cool before adding antibiotics.

17. Ampicillin (50 mg/ml solution in water).
18. LB agar plates (15 g/L) with 50 μg/ml ampicillin.
19. Sterile toothpicks.
20. Sterile polypropylene or polystyrene 14 ml culture tubes.
21. Column-based kits for DNA preparation (midiprep scale).
22. Column-based kits for DNA preparation (miniprep scale).

2.2. Growth and Inoculation of the Arabidopsis Plants

1. Plant growth cabinet.
2. Autoclave.
3. Plasmid pTY-S.
4. Positive-control plasmid pTY-PDS-IR.
5. *Arabidopsis thaliana* Col-0 seeds.
6. Peat pellets.
7. 90 mm square pots.
8. Soil mixture.
9. Watering trays.
10. Celite (or diatomaceous earth).
11. Aluminum foil.
12. Wash bottle.
13. Gloves.
14. Absorbent paper.

3. Methods

3.1. Design and Self-Hybridization of the Custom Oligonucleotide

To target a specific gene for silencing, we advise to design a self-hybridizing palindromic oligonucleotide of 80 nt (2 × 40 nt). Once inserted in the viral genome, this will generate a hairpin RNA bearing a stem of 38 nt and a 4 nt loop that is necessary and sufficient to promote silencing (8).

When designing the oligonucleotide, care has to be taken to select a target region that, once cloned into the *Sna*BI site of the pTY-S plasmid, will reconstitute an in-frame stop codon within the CP ORF (Fig. 1b). For that purpose, the 5′ end of the oligonucleotide should thus be either NTAG, NTAA, or NTGA, N being any of A, C, T, G (see Note 1).

1. Search the nucleotide sequence of the gene of interest for such motifs, which will constitute a suitable oligonucleotide 5′ end. The next 36 nucleotides should be identical to the target gene sequence. To obtain a palindromic oligonucleotide, this 40 nt sequence should then be duplicated in the reverse orientation to obtain an 80 nt sequence (Fig. 3).

```
Thioredoxin gene cDNA sequence

  1  TCTAAAACCCCTCAAAAGCTAAAACTGAAATCATTTTGAAGCAGAAATCG
 51  ATTTCCAAAATGGCTCTTGTTCAATCCAGAACTTTTCCACACCTCAACAC
101  TCCGTTATCGCCGATTCTATCCTCCCTCCATGCTCCGTCTTCTCTCTTTA
151  TCCGGAGGGAAATCCGACCGGTAGCCGCGCCGTTCTCCTCCTCCACCGCC
201  GGAAACTTGCCCTTCTCGCCGTTGACGCGTCCGCGAAAGCTTCTCTGTCC
251  TCCGCCTCGCGGCAAGTTTGTCAGAGAAGATTATCTTGTGAAGAAATTGC
301  AGCTCAAGAACTTCAGGAACTGGTGAAAGGAGATAGGAAGGTGCCGTTGA
351  TTGTTGATTTTTATGCGACATGGTGTGGACCTTGTATCTTGATGGCCCAG
401  GAACTAGAAATGCTTGCAGTAGAGTATGAGAGCAATGCAATAATCGTGAA
451  AGTTGATACAGATGATGAGTACGAGTTTGCACGCGACATGCAGGTTCGAG
501  GGTTACCTACATTATTCTTCATCAGTCCTGACCCAAGCAAAGATGCAATC
551  AGGACAGAAGGGCTAATTCCGTTACAGATGATGCACGATATCATTGACAA
601  CGAGATGTGAAGTTTTTACTGAGAAAACACTTTTGTACTCTTTATGATTT
651  GATTAGAATCGCCGTAATTTAGGATTGGTACCGTGTTGTTTGCATTTGAG
701  ATGTGGTAAGAATGTCTTCTAGTTTCATAGATGAAGAATGTAGTCTTAAG
751  ACATTTGTTTGTATGCAGTGTTCGTTTCAATACCTAAGTGAAAAGGTTAG
801  AGCTGTTGATCTGGTCGGATCCAAGTTATGTGGCACATCAAATTCACGAT
851  CCGGTCCATGTAATAACACTGTAGCTTAAACTTGGAAGAGAGAGATATGA
901  CGTCATT
```

To design an oligonucleotide, choose one of the motifs NTAA, NTAG or NTGA as a 5' end for instance
5' TTGA at position 222-225 of the gene sequence

Extend the sequence to the next 36 nucleotides (i.e. position 222-261 of the gene sequence)
5' TTGACGCGTCCGCGAAAGCTTCTCTGTCCTCCGCCTCGCG

Duplicate the 40-nt sequence in reverse orientation (shown in bold) to get a palindromic (inverted-repeat) sequence
5' TTGACGCGTCCGCGAAAGCTTCTCTGTCCTCCGCCTCGCG**CGCGAGGCGGAGGACAGAGAAGCTTTC GCGGACGCGTCAA** 3'

Fig. 3. Example of oligonucleotide design using the Arabidopsis *thioredoxin* gene (AT3G06730) as a target gene. Based on the cDNA nucleotide sequence shown below, 43 sequence motifs NTAG, NTAA, or NTGA can be identified (*gray boxes*), each of which could constitute the 5′ terminus of an oligonucleotide. Six of them (*underlined*) will reconstitute a *Sna*BI restriction site upon cloning and should be avoided (see Note 1).

2. Order the custom DNA oligonucleotide (0.025 μmol synthesis scale, desalted) from your favorite company.
3. Self-hybridize the custom DNA oligonucleotide by diluting 1 μg of oligonucleotide into 50 μl of hybridization buffer (10 mM Tris–HCl pH 7.5, 50 mM NaCl). Heat at 100°C for 5 min in a heating block, and let the block cool at room temperature (see Note 2).

3.2. Cloning of the Custom Oligonucleotide into the pTY-S Vector

1. Transform *E. coli* competent cells (DH5α strain) with the plasmid pTY-S and select transformants on LB agar plates with 50 μg/ml ampicillin incubated at 37°C for 16 h. Pick an individual colony using a sterile toothpick and inoculate a 14 ml culture tube containing 2.5 ml of LB with 50 μg/ml ampicillin. Allow cells to grow at 37°C for 6 h with vigorous shaking (200 rpm), then dilute this preculture in 100 ml of LB amp in a 1 L Erlenmeyer flask, and allow further growth at 37°C for 16 h. Prepare a glycerol stock of the transformed bacteria clone by mixing 0.7 ml of bacterial culture with 0.3 ml of sterile 100% glycerol and store at −80°C.

2. Harvest the remaining cells by centrifugation at 6,000×*g* for 20 min and extract the plasmid DNA of the bacterial culture using column-based kits for DNA preparation such as Qiagen or Macherey–Nagel midiprep plasmid kits. Determine the concentration of the plasmid DNA obtained by measuring the OD_{260} using a spectrophotometer or NanoDrop instrument.
3. Digest 5 μg of the plasmid pTY-S with the restriction enzyme *Sna*BI according to the supplier's instructions. To verify the proper linearization of the vector, analyze an aliquot (100 ng) of the digest by electrophoresis on a 0.8% agarose gel in TBE buffer containing ethidium bromide or Sybr Green, alongside DNA molecular weight markers (see Note 3). Visualize the DNA fragment under UV light using a transilluminator (see Note 4). The expected size of the linearized plasmid is 10,045 bp.
4. Inactivate the *Sna*BI restriction enzyme by treating the digestion mixture with 1 V of phenol/chloroform/isoamylic alcohol (see Note 5), and precipitate the linearized plasmid with 2 V of 96% ethanol. After storage at −20°C for at least 2 h, collect the precipitate by centrifugation at 13,000×*g* for 30 min at 4°C. Wash the DNA pellet with 70% ethanol and resuspend in 50 μl of water (see Note 6).
5. Perform ligation of the *Sna*BI-restricted pTY-S vector with the self-hybridized custom oligonucleotide (prepared in Subheading 3.1, step 3 described above) by mixing 100 ng (=1 μl) of plasmid DNA, with 100 ng (=5 μl) of oligonucleotide, 4 μl of 5× ligation buffer, and 1 μl of T4 DNA ligase in a final volume of 20 μl. Incubate overnight at 16°C (see Note 7).
6. To improve the efficiency of cloning (see Note 8), digest the ligation mixture by *Sna*BI by adding 2 μl of 10× restriction enzyme buffer and 1 μl of *Sna*BI restriction enzyme to the 20 μl ligation reaction. Incubate at 37°C for 1 h.
7. Use half of the reaction mixture to transform *E. coli* competent cells (DH5α strain) (see Note 9). Plate transformants on LB agar plates with 50 μg/ml ampicillin, and incubate at 30°C for 24 h (see Note 10). Pick clones individually using a sterile toothpick and inoculate 14 ml culture tubes containing 2.5 ml of LB with 50 μg/ml ampicillin. Allow cells to grow at 30°C for 16–24 h with vigorous shaking (200 rpm).
8. Harvest cells by spinning 1.2 ml at 6,000×*g* for 5 min and extract the plasmid DNA of each culture using standard alkaline lysis DNA extraction method (10) or column-based kits for DNA preparation such as Qiagen or Macherey–Nagel miniprep plasmid kits.
9. Screen the plasmid DNAs by digesting few microliters (200–500 ng) of each plasmid with a mixture of *Eco*RI and *Sma*I restriction enzymes according to the supplier's instructions.

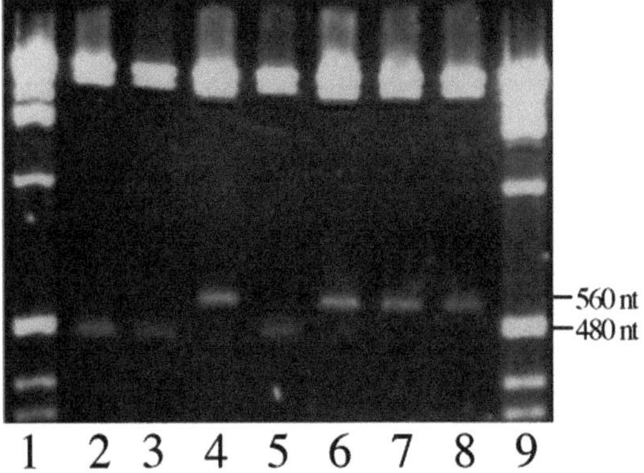

Fig. 4. Example of clone screening. DNA of pTY-S vector (lane 2) and that extracted from 6 bacterial colonies (lanes 3–8) was digested with EcoRI and SmaI and analyzed by electrophoresis on a 2% agarose gel run in TBE buffer. Lanes 1 and 9 correspond to the 1 Kb DNA ladder (Invitrogen). Four positive clones (in lanes 4 and 6–8) have integrated the 80 bp oligonucleotide as evidenced by the release of a 560 nt long DNA fragment.

Digest the pTY-S vector as a control, and analyze the resulting DNA fragments by electrophoresis on a 2% agarose gel in TBE buffer containing ethidium bromide or Sybr Green (see Note 3), alongside DNA molecular weight markers. Visualize the DNA fragment under UV light using a transilluminator (see Note 4). Digestion with *Eco*RI and *Sma*I should result in bands of 6,920, 2,637, and 480 nt for the control plasmid pTY-S or 6,920, 2,637, and 560 nt for the positive clones that have integrated the 80 mer oligonucleotide (Fig. 4).

10. Prepare a glycerol stock of the positive clones by mixing 0.7 ml of the bacterial culture with 0.3 ml of 100% glycerol and store at –80°C. Dilute the remaining bacterial culture into 10 ml of LB with 50 μg/ml ampicillin and allow cells to grow at 30°C for 24 h with vigorous shaking (200 rpm).
11. Harvest the cells by centrifugation at 6,000 × *g* 20 min and extract the plasmid DNA of the bacterial culture using column-based kits for DNA preparation such as Qiagen or Macherey–Nagel midiprep plasmid kits. Determine the concentration of the plasmid DNA obtained by measuring the OD_{260} using a spectrophotometer or NanoDrop instrument.
12. Verify the sequence of the positive clones using a primer that hybridizes upstream of the *Sna*BI cloning site (see Note 11). We recommend the use of particular sequencing methods for *difficult templates* that are adapted to read through inverted-repeat sequences (see Note 12) (Fig. 5).
13. Starting from the glycerol stock, streak the bacteria on LB agar plates with 50 μg/ml ampicillin, and incubate at 30°C for 24 h. Pick an individual colony using a sterile toothpick and

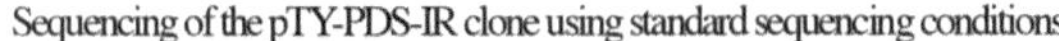

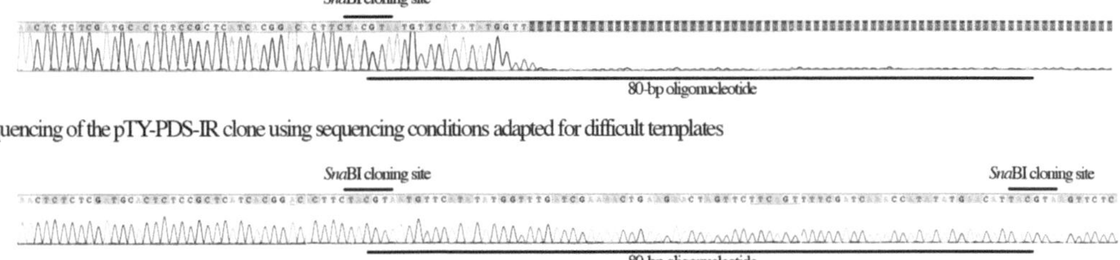

Fig. 5. Example of sequencing reaction of a positive clone. The plasmid pTY-PDS-IR bearing an 80 bp oligonucleotide has been sequenced using standard reaction conditions (*top*) or conditions adapted for difficult templates (*bottom*), in which case the inverted-repeat sequence can readily be verified. The sequencing reactions have been performed by the company Cogenics.

grow the bacterial culture in LB with 50 μg/ml ampicillin at 30°C. Isolate DNA in sufficient amounts for the number of plants to be inoculated for each construct (see Note 13).

3.3. Growth and Inoculation of the Arabidopsis Plants

When designing a VIGS experiment, make sure to include all of the proper controls. We recommend including mock (= water) inoculation, the *empty* VIGS vector inoculation (pTY-S), a VIGS vector containing an 80 nt oligonucleotide unrelated to the target gene as a control for viral symptoms, a VIGS vector with a visible phenotype such as pTY-PDS-IR which causes photobleaching because gene silencing of the *phytoene desaturase* gene, and the VIGS vector containing the 80 nt oligonucleotide related to the target gene. We routinely inoculate three plants of each control and six plants with the experimental vector.

Appropriate biosafety precautions need to be practiced to prevent unintentional escape of the VIGS vector into the environment:

1. Germinate *Arabidopsis thaliana* Col-0 seeds (see Note 14) in peat pellets at 20–22°C under a 16 h/8 h light/dark cycle in a growth chamber. Keep the peat moist at all times. Transplant 2 week old seedlings into individual soil pots inserted in a watering tray (see Note 15). Plants at the 14 leaf stage, before the occurrence of bolting, are optimal for inoculation (Fig. 6a) (see Note 16). Place the plants in the dark (i.e., in a cabinet) 16–24 h prior to inoculation.
2. Cover the celite pot with an aluminum foil pierced with small holes, to use it as a *salt-shaker* (Fig. 7) and lightly dust the three main leaves of each Arabidopsis plants with celite (see Note 17). The optimal amount of abrasive is shown in Fig. 6b (see Note 18).
3. Prepare the inoculum, consisting in 3 μg column-purified DNA plasmid, diluted in 25 μl of water, per plant to be inoculated. Using a micropipette, drop 8 μl of plasmid solution onto

Fig. 6. Arabidopsis growth stage for inoculation. Arabidopsis plant at the optimal growth stage for inoculation, before (**a**) and after (**b**) the abrasive has been dusted.

Fig. 7. Use of celite as an abrasive. The pot of celite, covered with an aluminum foil pierced with small holes, is used as a *salt-shaker* to lightly dust the Arabidopsis leaves.

the first leaf to be inoculated (Fig. 8a). Wearing gloves, gently rub the liquid across the surface of the leaf three to four times using one finger while maintaining the leaf with one finger of the other hand (Fig. 8b, c). Do the same with the two other leaves to be inoculated (see Notes 19 and 20). Change gloves between each plant.

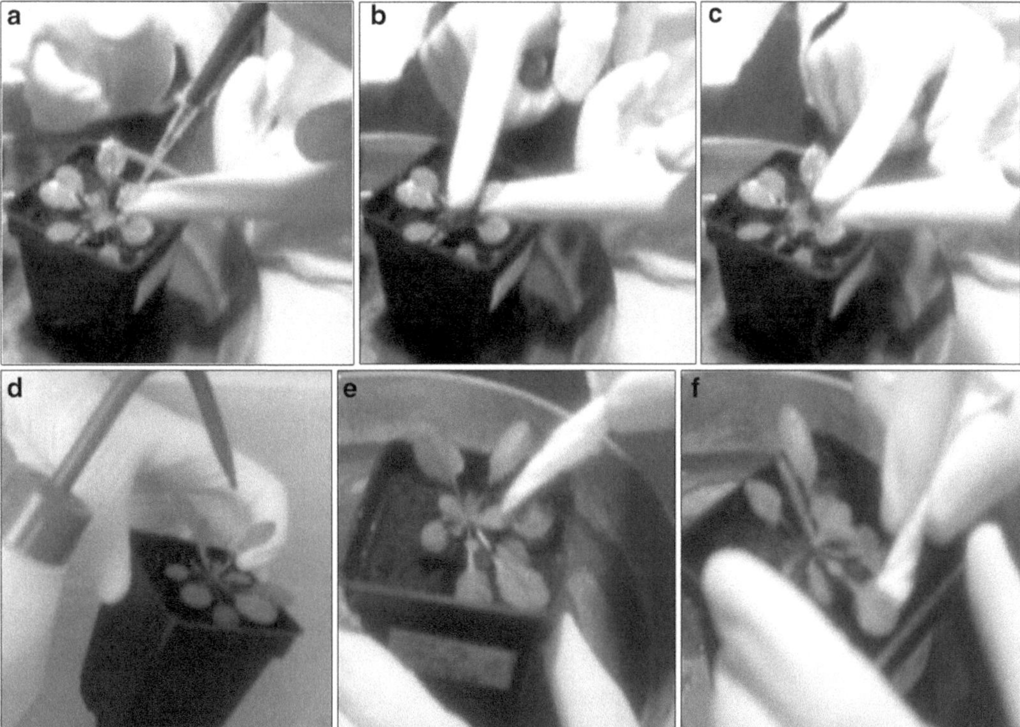

Fig. 8. Different steps in Arabidopsis inoculation. A drop of plasmid solution (8 μl) is deposited on the leaf to be inoculated (**a**) and gently rubbed 3–4 times over the surface of the leaf with a gloved finger (**b**, **c**). The inoculum is then washed off using a wash bottle (**d**), and the excess water on the upper and lower sides of each leaves is removed using absorbent paper (**e**, **f**).

4. Wait for 1–2 min (see Note 21) and then wash off each leaf for 15 s using a wash bottle filled with tap water (Fig. 8d). Work over a watering tray to collect water effluents (see Note 22).
5. Remove the excess water from the leaves (both the upper and lower sides) using absorbent paper (Fig. 8e, f) (see Note 23).
6. Maintain the inoculated plants at 20–22°C under a 16 h/8 h light/dark cycle in a growth chamber (see Note 24). Photobleaching of leaves and stem resulting from silencing of the *phytoene desaturase* gene should be visible in the control plant inoculated with pTY-PDS-IR about 3 weeks post-inoculation (Fig. 2) (see Note 25). Once silencing is initiated, it persists throughout the life of the plant.
7. Confirmation of VIGS of a candidate gene and determination of the extent of down-regulation should be performed by semiquantitative RT-PCR or qRT-PCR using a primer pair annealing outside the region targeted for silencing, so that only the endogenous mRNA is probed.

4. Notes

1. It is also strongly advisable to select sequences that once cloned into the pTY-S vector will not reconstitute the *Sna*BI restriction site (TAC/GTA) because, in this situation, the desired clones can be counterselected by a *Sna*BI digest, which facilitates the cloning step. Therefore, it is better to avoid oligonucleotides with a 5′ end being either GTAG or GTAA.
2. Self-hybridization of the oligonucleotide can also be performed using a PCR machine.
3. Ethidium bromide is a known mutagen that should be handled and disposed in accordance with appropriate safety regulations.
4. Use protective gear to limit exposure to ultraviolet radiation.
5. Phenol/chloroform is toxic and should be handled and disposed in accordance with appropriate safety regulations.
6. It is suitable to analyze an aliquot (1 μl) of the linearized plasmid by electrophoresis on a 0.8% agarose gel to verify the DNA has not been lost during ethanol precipitation and wash.
7. In the case where a refrigerated water bath is not available, place a regular water bath in a cold room.
8. This counterselection step should be done only in the case where the inserted oligonucleotide does not reconstitute a *Sna*BI restriction site. The plasmid constructs that have integrated the oligonucleotide will be insensitive to the digest and remain circular. They will transform *E. coli* with a greater efficiency than the empty plasmid vector that will be digested by *Sna*BI, as linearized DNA transforms bacteria with less efficiency than circular DNA.
9. We routinely use thermocompetent bacteria with transformation efficiency of 10^6 cfu/μg.
10. We observed that the recombinant plasmids may be toxic to the bacterial cells presumably because of the hairpin structure that may form during plasmid replication. We therefore recommend growing the bacterial cells at 30°C instead of 37°C.
11. We routinely use the oligonucleotide 5′ GGCGCCATCAACA CCCTCTCACC 3′ to sequence the pTY-S-derived plasmids as it hybridizes about 180 nt upstream of the *Sna*BI cloning site.
12. If the sequencing reaction is performed using standard conditions, the sequence will fail and stop within the inverted repeat because of its high stability, unless the sequence contains a point mutation, that causes destabilization of the hairpin. Therefore, failure of the sequencing reaction using standard conditions is a good indication that the sequence of the clone is good!

13. One microgram of plasmid per plant was shown to be sufficient to achieve 100% infection, but we routinely use 3 μg per plant.
14. As severe viral symptoms were observed upon infection of WS and Ler ecotypes, we do not recommend the use of pTY-S-derived vectors on these ecotypes.
15. The sanitary status of the plant is important, and excessive watering of the plants should be avoided to prevent pest infections.
16. If a short stem is present in some plants, it can be cut prior to inoculation.
17. Celite is dangerous to breathe and a dust mask is recommended when working with it.
18. The petioles of the leaves to be inoculated can be marked with a marker pen, but care has to be taken not to damage them.
19. The inoculation is easier if performed by two persons: one who is pipetting the plasmid solution and the other one who is rubbing the leaves.
20. Do not rub the leaf more than 3–4 times, as it can be damaged by the abrasive.
21. Do not let the leaves dry before washing them, as this will cause damage. It is thus advisable not to inoculate more than two to three plants at once.
22. Appropriate biosafety precautions need to be practiced to prevent unintentional escape of the VIGS vector into the environment. In particular, waste disposal should be taken care according to local regulations regarding GMOs.
23. Do not let the leaves stick to the soil or to the plastic of the pot, as this will damage them.
24. Temperature affects the efficiency of viral infection as well as VIGS efficiency. Robust silencing results were obtained when plants were grown at 20–22°C.
25. The precise timing of the VIGS is very dependent on environmental conditions, and it is advisable to set up tests with the pTY-PDS-IR control vector to monitor the progress of the infection and silencing processes.

References

1. Martinez-Herrera D, Romero J, Martinez-Zapater JM et al (1994) Suitability of *Arabidopsis thaliana* as a system for the study of plant-virus interactions. Fitopatologia 29: 132–136
2. Matthews REF (1973) Induction of disease by viruses, with special reference to turnip yellow mosaic virus. Annu Rev Phytopathol 11: 147–170
3. Bozarth CS, Weiland JJ, Dreher TW (1992) Expression of ORF-69 of turnip yellow mosaic virus is necessary for viral spread in plants. Virology 187:124–130
4. Chen J, Li WX, Xie D et al (2004) Viral virulence protein suppresses RNA silencing—mediated defense but upregulates the role of microRNA in host gene expression. Plant Cell 16:1302–1313

5. Jakubiec A, Drugeon G, Camborde L et al (2007) Proteolytic processing of Turnip yellow mosaic virus replication proteins and functional impact on infectivity. J Virol 81:11402–11412
6. Weiland JJ, Dreher TW (1989) Infectious TYMV RNA from cloned cDNA: effects in vitro and in vivo of point substitutions in the initiation codons of two extensively overlapping ORFs. Nucleic Acids Res 17: 4675–4687
7. Schirawski J, Voyatzakis A, Zaccomer B et al (2000) Identification and functional analysis of the turnip yellow mosaic tymovirus subgenomic promoter. J Virol 74:11073–11080
8. Pflieger S, Blanchet S, Camborde L et al (2008) Efficient virus-induced gene silencing in Arabidopsis using a 'one-step' TYMV-derived vector. Plant J 56:678–690
9. Voinnet O (2005) Induction and suppression of RNA silencing: insights from viral infections. Nat Rev Genet 6:206–220
10. Birnboim HC, Doly J (1979) A rapid alkaline extraction procedure for screening recombinant plasmid DNA. Nucleic Acids Res 7:1513–1523

Chapter 16

Virus-Induced Gene Silencing in Strawberry Fruit

Haifeng Jia and Yuanyue Shen

Abstract

Virus-induced gene silencing (VIGS) is a technology that exploits an RNA-mediated antiviral defense mechanism and which has great potential for use in plant reverse genetics. Recently, whole-genome studies and gene sequencing in plants have produced a massive amount of sequence information. A major challenge for plant biologists is to convert this sequence information into functional information. In this study, we demonstrate that VIGS can be used to determine gene functions in strawberry and that it is a powerful new tool for studying fruit ripening. The ABA synthetic gene *FaNCED1*, which can promote strawberry fruit ripening, was used as the reporter gene. In this chapter, we describe the use of TRV-mediated VIGS in strawberry fruit.

Key word: Virus-induced gene silencing (VIGS)

1. Introduction

The antisense-mediated inhibition of gene expression is commonly used to downregulate gene expression in plants (1–4); however, this approach relies on stable transformation, which is difficult in many plant species. In addition, gene loss-of-function can result in death early in plant development. Moreover, transformation has other disadvantages such as the complexity of the procedure, the labor-intensive screening, and the low efficiency (5–7). Fortunately, virus-induced gene silencing (VIGS) offers an attractive and quick alternative to knocking out the expression of a gene of interest, and it avoids the need for transformation (8–10). This method has several other advantages over stable transformation. First, constructs can be easily assembled by inserting a fragment of the target gene into a virus vector. Second, constructs can be assembled in a short period of time, and the VIGS phenotype can be observed

Annette Becker (ed.), *Virus-Induced Gene Silencing: Methods and Protocols*, Methods in Molecular Biology, vol. 975, DOI 10.1007/978-1-62703-278-0_16,

shortly after inoculation. Consequently, it is feasible to perform a high-throughput analysis of many genes. Third, VIGS can be used to study gene families. Unlike stable transformation procedures, VIGS also permits the analysis of genes with a lethal phenotype in antisense knockout plants (6, 10, 11).

Many VIGS vectors have been developed using RNA viruses such as tobacco rattle virus (TRV) (12, 13), potato virus X (14), barley stripe mosaic virus (15), and DNA viruses such as TGMV (16) and CaLCuV (17). Because most plant viruses are unable to invade meristematic tissue, it is difficult to silence genes in reproductive organs such as flowers and fruits. However, it has recently been shown that TRV can invade meristematic tissue and induce gene silencing in petunia and tobacco flowers (18, 19). Although VIGS has been applied to roots (20, 21) and tubers (22), it has mainly been used on leaves and in defensive signal transduction (12, 23). It has also been used to functionally characterize genes in tobacco (12, 23), *Arabidopsis* (18), tomato (12), barley (15), pepper (24), potato (22, 25), legumes (26), cassava (27), petunia (18), peach (28), and strawberry (29).

Several genes expressed during fruit ripening have been isolated using classical differential display and DNA microarrays (30–32). With the sequencing of expressed sequence tags, the number of ripening-related genes has increased rapidly, and there is a pressing need for large-scale functional gene identification. In this study, we demonstrate that TRV-based VIGS is effective in triggering VIGS in strawberry fruit. When the virus infects and spreads systemically throughout a plant, the endogenous gene transcripts, which are homologous to the inserted fragment in the viral vector (VIGS-vector), are degraded by posttranscriptional gene silencing (PTGS) (5). PTGS is a plant defense mechanism that results in the specific degradation of a population of homologous transgenes encoding RNA or virus RNA (7).We also systemically silenced the strawberry 9-cis-epoxycarotenoid dioxygenase gene (*FaNCED1*), which plays a role in ABA synthesis. This approach will enable the large-scale functional analysis of genes involved in strawberry fruit ripening.

2. Materials and Methods

2.1. Plant Material and Growth Conditions

Octaploid strawberry (*Fragaria* × *ananassa* cv. Fugilia) shall be used in all experiments. Standard growth conditions should be maintained in a greenhouse under standard culture conditions (25/15°C, 16/8 h day/night, relative humidity (RH) 70–90%). Approximately 60 flowers on 10 strawberry plants are tagged on the day of anthesis. Three sampling replications are recommended.

The receptacle seeds have to be removed and cut into small cubes (0.5–0.8 cm^3) then frozen quickly in liquid nitrogen and stored at −80°C.

2.2. Agrobacterium Strains

Agrobacterium strain GV3101 containing pTRV1 or pTRV2 and its derivatives pTRV2-target gene are used for VIGS. *Agrobacterium* strain GV3101 containing the TRV-VIGS vectors is grown at 28°C in LB medium containing 10 mM MES and 20 μM acetosyringone with appropriate antibiotics. After 24 h, the cells are harvested and resuspended in *Agrobacterium* infiltration buffer (below) to a final OD_{600} of 1.0 (for pTRV1 or pTRV2 and its derivatives) and shaken for 4 h at room temperature before infiltration.

2.3. Buffers for Bacteria and Infiltration of Plants

The following buffers and primers are prepared:

1. To make LB liquid medium, dissolve 10 g of tryptone, 5 g of yeast extract, 10 g of NaCl, and 15 g of agar (for LB solid medium) in 1 L of molecular biology grade water. Adjust the pH to 7.0 and sterilize the mixture by autoclaving.
2. To make 0.15 M NaCl, dissolve 8.766 g of NaCl in 1 L of deionized water.
3. To make 50 mM $CaCl_2$, dissolve 5.549 g of $CaCl_2$ in 1 L of deionized water.
4. The infection buffer contains 10 mM $MgCl_2$, 10 mM MES (pH 5.6), and 200 μM acetosyringone.
5. To make buffer A, resuspend *Agrobacterium* cells containing pTRV1 or pTRV2 in infection buffer and mixed (1:1 (v/v) volume).
6. To make buffer B, resuspend *Agrobacterium* cells containing pTRV1- or pTRV2-target gene in infection buffer and mixed (1:1 (v/v) volume). When the infection material has been prepared, the *Agrobacterium* suspension is evenly injected throughout the entire fruit while it is still attached to the plant about 14 days after pollination using a sterile 1 ml hypodermic syringe.
7. Random hexamer primers are used for cDNA synthesis to detect the RNA1 or RNA2 and oligo (dT)18 was used for all other cDNA syntheses. Prepare random primers (for the detection of RNA1 or RNA2) using the sequence 5′-NNNNNN-3′ (N represents A, T, G, and C), and for preparation of oligo (dT) 18, use the sequence 5′-TTTTTTTTTTTTTTTTTT-3′ for cDNA.

3. Methods

3.1. Preparation of Competent Agrobacterium tumefaciens

All procedures are carried out on a clean bench.

1. Monoclonal GV3101 strains are used to inoculate 2 mL of LB liquid medium and incubated overnight at 28°C.
2. The overnight culture is used to inoculate 50 mL of LB liquid medium, which is incubated at 28°C with shaking at 200 rpm, and cultured to $OD_{600} = 0.5$.
3. The cells are centrifuged at 2,800 *g* for 5 min and resuspended in 10 mL of 0.15 M NaCl.
4. The cells are centrifuged at 2,800 *g* for 5 min and resuspended in 1 mL of precooled 50 mM $CaCl_2$.
5. Following the addition of 20% glycerol (v/v), the competent cells are snap frozen in liquid nitrogen for 1 min and stored at -70°C.

3.2. Preparation of Agrobacteria Containing the VIGS Vectors

To silence a target gene in strawberry, a gene-specific fragment of a cDNA (refer to Subheading 3.3, step 7) of approximately 650 bp in length (Fig. 1) is PCR amplified with appropriate restriction sites (corresponding to the TRV pTRV2 restriction sites) flanking the target gene sequence. Subsequently, the target gene fragment is inserted into the pTRV2 restriction sites.

PCR Amplification, Purification

PCR is performed under the following conditions: 94°C for 5 min, followed by 35 cycles at 94°C for 30 s, 53°C for 40 s, and 72°C for 1 min, followed by a final incubation at 72°C for 10 min. The PCR products are purified with a PCR Purification Kit (Takara Bio), ligated into a pMD19-T vector, and subsequently transformed into *Escherichia coli* DH5a to sequence.

Restriction Digest System

The TRV2 vector and the fragment are both cut with the appropriate restriction endonuclease and buffer at 37°C for 3 h; they are purified with a PCR Purification Kit (Takara Bio) to prepare for the ligation.

Ligation

The ligation reaction contains appropriate amounts of digested pTRV2 vector, an equally digested target gene fragment, 1 μl DNA T4 ligase, and 2 μl 10× buffer in 20 μl total volume. All reaction components are mixed thoroughly and incubated in 5 ml liquid LB containing the appropriate antibiotic at 16°C overnight. The ligation mix is then transformed into competent *Escherichia coli* DH5a cells using standard protocols to obtain a colony that contains pTRV2 with the inserted target gene sequence.

Transformation of the pTRV Vectors into *A. tumefaciens*

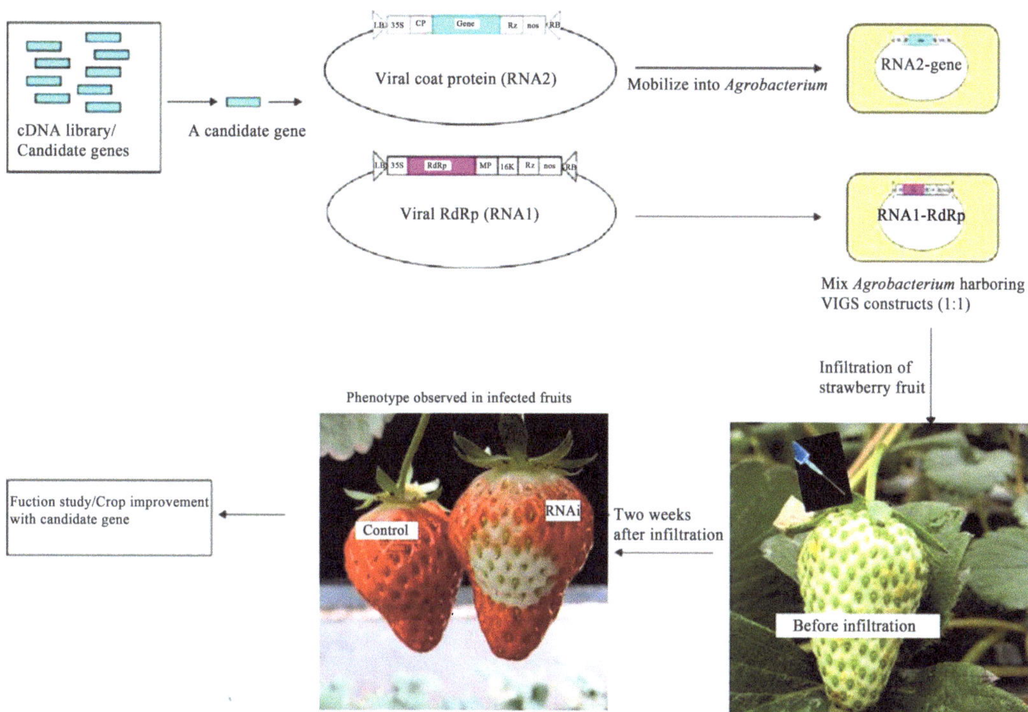

Fig. 1. TRV is a double-stranded RNA virus composed of two single-stranded RNAs: RNA1 and RNA2. RNA1 encodes, among others, the RNA-dependent RNA polymerase, while RNA2 encodes the coat protein and other important proteins. Target gene fragments were inserted into the RNA2 sequence. In our Agrobacterium infection method, the modified viral genome is transferred to strawberry fruits. After infection, mixtures of strawberry fruit (VIGS treated and control) were collected for expression analysis. Silencing phenotypes appeared 2 weeks after infection. Thus, combined with the infectivity of Agrobacterium, VIGS is a convenient infection method for strawberry fruit.

1. pTRV1, pTRV2-E (empty vector), and the recombinant vector pTRV2-target gene are transformed into *Agrobacterium* cells.
2. The competent cells (200 μL) are mixed with 1 μg of vector DNA and incubated in an ice bath for 30 min.
3. The cells are then incubated in liquid nitrogen for 3–5 min, followed by incubation at 37°C for 5 min.
4. The cells are suspended in LB liquid medium (1 mL), incubated at 28°C for 4 h, and centrifuged at 11,190 *g* for 30 s.
5. The supernatant is discarded, and 200 μL of LB liquid medium is added to the cells. The cells are then added to an LB-agar plate with the appropriate antibiotics and incubated at 28°C for 2 days.

3.3. Infection of Strawberriesl with the Agrobacterium Strains

1. *Agrobacterium* cells containing pTRV1, pTRV2-E, or pTRV2-target gene are cultured in LB liquid medium overnight at 28°C to an OD_{600} of 1.0–2.0. The cells are then collected, and

their OD_{600} is adjusted to between 1.0 and 2.0 with infection buffer.

2. The *Agrobacteria* containing the pTRV vectors in infection buffer are incubated at 28°C for 4 h before strawberry fruit infection.
3. *Agrobacteria* containing the pTRV1 and pTRV2-E vectors were mixed in equal volumes to the control (buffer A). pTRV1 and pTRV2-target gene vectors were mixed in equal volumes to silence the target gene (buffer B).
4. Strawberry fruit attached to plants approximately 14 days after pollination is selected. Buffers A and B are injected into the entire fruit with a sterile 1 mL syringe attached to a needle, respectively (Fig. 1).
5. 2 weeks after inoculation, phenotypic changes are observed. The RNAi-silenced and control fruit is evaluated for both TRV and the target gene (Fig. 1).
6. Total RNA is extracted from 10 g of fresh untreated or treated strawberry fruit using an RNA extraction kit (SV Total RNA Isolation System: Promega). Genomic DNA is eliminated with a 15 min incubation at 37°C with RNase-Free DNase (Takara Bio), followed by purification with an RNA Clean Purification Kit (BioTeke).

   The purity and integrity of the RNA is determined by agarose gel electrophoresis and spectrophotometry. High-quality RNA is used for Northern blotting and RT-PCR.
7. To generate first-strand cDNA, 1 μg of total RNA was reverse transcribed using a Clontech Kit (Takara Bio) according to the manufacturer's protocol. The protocol was as follows:

   1 μg RNA and 1 μl (50 μM) random (or oligo d(T)18) are mixed and incubated at 70°C for 10 min, centrifuged briefly, and placed back on ice, then 4 μl 5× reverse buffer, 2 μl dNTPs (10 μM), 1 μl MMLV reverse transcriptase, and 0.5 μl RNAase inhibitor are added and mixed. Finally, RNAase-free water is added to 20 μl. The reaction mix is incubated at 42°C for 1 h followed by a 70°C for 15 min and stored at −20°C for further use.
8. The following strawberry-specific and TRV-specific primers can be used:

   Actin(GenBank accession number AB116565)

   Forward:5′— TGGGTTTGCTGGAGATGAT—3′; Reverse:5′—CAGTAGGAGAACTGGGTGC—3′

   NCED1(GenBank accession number HQ290318)

Forward:5′—ACGACTTCGCCATTACCG—3′; Reverse:5′—AGCATCGCTCGCATTCT—3′RNA1(GenBank accession number AF406990)

Forward:5′—TTACAGGTTATTTGGGCTAG—3′; Reverse:5′—CCGGGTTCAATTCCTTATC—3′

RNA2(GenBank accession number AF406991)

Forward:5′—GTTAGATAATGGTTTGGTGGTC—3′; Reverse:5′—GCGTTCTCCCGTTTCGTC

Acknowledgments

We would like to thank Prof. Liu Yu-le (Qinghua University, Beijing, China) for pTRV vectors. This work was supported financially by the National Key Basic Research '973' Program of China (grant No. 2012CB126306), the National Science Foundation of China (grant No. 30971977, 31272144), and the Newstar of Science and Technology Supported by Beijing Metropolis (grant No. Z111105054511048).

References

1. Gray J, Picton S, Shabbeer J et al (1992) Molecular biology of fruit ripening and its manipulation with antisense genes. Plant Mol Biol 19:69–87
2. Hedden P, Phillips AL (2000) Manipulation of hormone biosynthetic genes in transgenic plants. Curr Opin Biotech 11:130–137
3. Herrera-Estrella L, Simpson J, Martinez-Trujillo M (2005) Transgenic plants: an historical perspective. Methods Mol Biol 286:3–32
4. Stearns JC, Glick BR (2003) Transgenic plants with altered ethylene biosynthesis or perception. Biotechnol Adv 21:193–210
5. Baulcombe DC (2004) RNA silencing in plants. Nature 431:356–363
6. Benedito VA, Visser PB, Angenent GC et al (2004) The potential of virus-induced gene silencing for speeding up functional characterization of plant genes. Genet Mol Res 3:323–341
7. Burch-Smith TM, Anderson JC, Martin GB et al (2004) Applications and advantages of virus-induced gene silencing for gene function studies in plants. Plant J 39:734–746
8. Baulcombe DC (1999) Fast forward genetics based on virus induced gene silencing. Curr Opin Plant Biol 2:109–113
9. Dinesh-Kumar SP, Anandalakshmi R, Marathe R et al (2003) Virus-induced gene silencing. Methods Mol Biol 236:287–294
10. Fu DQ, Zhu BZ, Zhao XD et al (2005) Advance of the virus-induced gene silencing in plant. J Chinese Biotech 1:89–95
11. Robertson D (2004) VIGS vectors for gene silencing: many targets, many tools. Annu Rev Plant Biol 55:495–519
12. Liu Y, Schiff M, Dinesh-Kumar SP (2002) Virus-induced gene silencing in tomato. Plant J 31:777–786
13. Ratcliff F, Martin-Hernandez AM, Baulcombe DC (2001) Tobacco rattle virus as a vector for analysis of gene function by silencing. Plant J 25:237–245
14. Angell SM, Baulcombe DC (1999) Potato virus X amplicon-mediated silencing of nuclear genes. Plant J 20:357–362
15. Holzberg S, Brosio P, Gross C et al (2002) Barley stripe mosaic virus-induced gene silencing in a monocot plant. Plant J 30:315–327
16. Peele C, Jordan CV, Muangsan N et al (2001) Silencing of a meristematic gene using geminivirus-derived vectors. Plant J 27:357–366

17. Turnage MA, Muangsan N, Peele CG et al (2002) Geminivirus-based vectors for gene silencing in Arabidopsis. Plant J 30:107–114
18. Chen JC, Jiang CZ, Gookin TE et al (2004) Chalcone synthase as a reporter in virus-induced gene silencing studies of flower senescence. Plant Mol Biol 55:521–530
19. Liu Y, Nakayama N, Schiff M et al (2004) Virus induced gene silencing of a DEFICIENS ortholog in Nicotiana benthamiana. Plant Mol Biol 54:701–711
20. Ryu CM, Anand A, Kang L et al (2004) Agrodrench: a novel and effective agroinoculation method for virus-induced gene silencing in roots and diverse Solanaceous species. Plant J 40:322–331
21. Valentine T, Shaw J, Blok VC et al (2004) Efficient virus-induced gene silencing in roots using a modified tobacco rattle virus vector. Plant Physiol 136:3999–4009
22. Faivre-Rampant O, Gilroy EM, Hrubikova K et al (2004) Potato virus X-induced gene silencing in leaves and tubers of potato. Plant Physiol 134:1308–1316
23. Peart JR, Cook G, Feys BJ et al (2002) An EDS1 orthologue is required for N-mediated resistance against tobacco mosaic virus. Plant J 29:569–579
24. Chung E, Seong E, Kim YC et al (2004) A method of high frequency virus-induced gene silencing in chili pepper (Capsicum annuum L. cv. Bukang). Mol Cell 17:377–380
25. Brigneti G, Martin-Hernandez AM, Jin H et al (2004) Virus-induced gene silencing in Solanum species. Plant J 39:264–272
26. Constantin GD, Krath BN, Macfarlane SA et al (2004) Virus-induced gene silencing as a tool for functional genomics in a legume species. Plant J 40:622–631
27. Fofana IB, Sangare A, Collier R et al (2004) A geminivirus-induced gene silencing system for gene function validation in cassava. Plant Mol Biol 56:613–624
28. Jia H-F, Chai Y-M, Shen Y-Y (2011) Cloning and Characterization of the H Subunit of a Magnesium Chelatase Gene (PpCHLH) in Peach. J Plant Growth Regul 30:445–455
29. Jia H-F, Chai Y-M, Li C-L et al (2011) Abscisic Acid Plays an Important Role in the Regulation of Strawberry Fruit Ripening. Plant Physiol 157:188–199
30. Moore S, Vrebalov J, Payton P et al (2002) Use of genomics tools to isolate key ripening genes and analyse fruit maturation in tomato. J Exp Bot 53:2023–2030
31. Zegzouti H, Jones B, Frasse P et al (1999) Ethylene-regulated gene expression in tomato fruit: characterization of novel ethylene-responsive and ripening-related genes isolated by differential display. Plant J 18:589–600
32. Aharoni A, O'Connell AP (2002) Gene expression analysis of strawberry achene and receptacle maturation using DNA microarrays. J Exp Bot 53:2073–2087

Index

Annette Becker (ed.), *Virus-Induced Gene Silencing: Methods and Protocols*, Methods in Molecular Biology, vol. 975, DOI 10.1007/978-1-62703-278-0,

If you have any concerns about our products,
you can contact us on
ProductSafety@springernature.com

In case Publisher is established outside the EU,
the EU authorized representative is:
Springer Nature Customer Service Center GmbH
Europaplatz 3, 69115 Heidelberg, Germany

Printed by Libri Plureos GmbH
in Hamburg, Germany